T0269076

COMPUTATIONAL ALGEBRAIC GEOMETRY

LONDON MATHEMATICAL SOCIETY STUDENT TEXTS

Managing editor: Professor J.W. Bruce, Department of Mathematics,
University of Liverpool, UK

COMPUTATIONAL ALGEBRAIC GEOMETRY

HAL SCHENCK

Texas A&M University

CAMBRIDGE UNIVERSITY PRESS
Cambridge, New York, Melbourne, Madrid, Cape Town, Singapore,
São Paulo, Delhi, Dubai, Tokyo, Mexico City

Cambridge University Press
The Edinburgh Building, Cambridge CB2 8RU, UK

Published in the United States of America by Cambridge University Press, New York

www.cambridge.org
Information on this title: www.cambridge.org/9780521536509

First published 2003

A catalogue record for this publication is available from the British Library

Library of Congress Cataloguing in Publication Data
Schenck, Hal.
Computational algebraic geometry / Hal Schenck.
p.   cm.- (London Mathematical Society student texts ; 58)
Includes bibliographical references and index.
ISBN 0-521-82964-X (hardback) - ISBN 0-521-53650-2 (pbk.)
I. Geometry, Algebraic - Data processing - Congresses.   I. Title.   II. Series
QA564.S29   2003
516.3'5 - dc21   2003053074

ISBN 978-0-521-82964-9 Hardback
ISBN 978-0-521-53650-9 Paperback

To Mom and Dad

Contents

Preface

Although the title of this book is "Computational Algebraic Geometry", it could also be titled "Snapshots of Commutative Algebra via Macaulay 2". The aim is to bring algebra, geometry, and combinatorics to life by examining the interplay between these areas; it also provides the reader with a taste of algebra different from the usual beginning graduate student diet of groups and field theory. As background the prerequisite is a decent grounding in abstract algebra at the level of [56]; familiarity with some topology and complex analysis would be nice but is not indispensable. The snapshots which are included here come from commutative algebra, algebraic geometry, algebraic topology, and algebraic combinatorics. All are set against a backdrop of homological algebra. There are several reasons for this: first and foremost, homological algebra is the common thread which ties everything together. The second reason is that many computational techniques involve homological algebra in a fundamental way; for example, a recurring motif is the idea of replacing a complicated object with a sequence of simple objects. The last reason is personal – I wanted to give the staid and abstract constructs of homological algebra (e.g. derived functors) a chance to get out and strut their stuff. This is said only half jokingly – in the first class I ever had in homological algebra, I asked the professor what good Tor was; the answer that Tor is the derived functor of tensor product did not grip me. When I complained to my advisor, he said "Ah, but you can give a two line proof of the Hilbert syzygy theorem using Tor – go figure it out". What an epiphany it was! Note to student: if you don't know what homological algebra and derived functors are, one point of this book is to give a hands-on introduction to these topics.

Of course, to understand anything means being able to compute examples, so oftentimes rather than dwelling on details best left to specialized texts (e.g. showing simplicial homology is indeed a topological invariant) we plunge blithely forward into computations (both by hand and by computer) in order to get a feel for how things work. This engineering mentality may be bothersome to the fastidious reader, but the first word in the title is not "Theoretical" but "Computational". We work mostly in the category of graded rings and modules, so the geometric setting is usually projective space. One unifying theme is the study of finite free resolutions; in particular, lots of the geometric invariants we study can be read off from a free resolution. Advances in computing and algorithms over the last twenty years mean that these gadgets are actually computable, so we can get our hands dirty doing lots of examples. By the end of the book the reader should feel comfortable talking about the degree and genus of a curve, the dimension and Hilbert polynomial of a variety, the Stanley–Reisner ring of a simplicial complex (and simplicial homology) and such abstract things as Ext, Tor, and regularity. Overall, the book is something of an algebra smorgasbord, moving from an appetizer of commutative algebra to homological methods. Of course, homological algebra would be less tasty without a garnish of history, so we add a dash of algebraic topology and a pinch of simplicial complexes and combinatorics. For dessert, we give Stanley's beautiful application of these methods to solve a combinatorial problem (the upper bound conjecture for spheres).

One of the wonderful things about computational algebra is that it is very easy to generate and test ideas. There are numerous exercises where the reader is asked to write scripts to test open research conjectures; the idea is to get folks thinking about open problems at an early stage. It is also exciting to find (albeit a couple years too late!) a counterexample to a published conjecture; the reader gets a chance to do this. In short, the exercises are geared at convincing students that doing research mathematics does not consist solely of ruminating alone in a darkened room, but also of rolling up one's sleeves, writing some code, and having the computer do the legwork.

Rather than giving examples of scripts in pseudocode, I have chosen to use a specific computer algebra package (Macaulay 2, by Dan Grayson and Mike Stillman). Macaulay 2 is free, easy to use, fast and flexible. Another virtue of Macaulay 2 is that the syntax is pretty straightforward. Thus, Macaulay 2 scripts *look like* pseudocode, but the reader can have the satisfaction of typing in scripts and seeing them run. Macaulay 2 works over finite fields of characteristic ≤ 32749, also over $\mathbb{Q}$ and certain other fields of characteristic zero. The examples in this book are often computed over finite fields. As Eisenbud notes in [32] "Experience with the sort of computation we will be doing shows that working over $\mathbb{Z}/p$, where p is a moderately large prime, gives results identical to the results we would get in characteristic 0".

I include here a mea culpa. This book grew from a dilemma – to give students a tapa of advanced algebra means that one would like to include snippets from

commutative algebra	algebraic geometry	topology and combinatorics
Atiyah–Macdonald [3]	*Cox–Little–O'Shea* [23]	*Fulton* [41]
Balcerzyk–Jozefiak [6]	*Griffiths* [48]	*Munkres* [71]
Bruns–Herzog [21]	*Harris* [52]	*Spanier* [87]
Eisenbud [28]	*Hartshorne* [53]	*Stanley* [88]
Matsumura [64]	*Miranda* [69]	*Sturmfels* [92]
Sharp [84]	*Reid* [78]	*Weibel* [98]
Vasconcelos [95]	*Shafarevich* [82]	*Ziegler* [100]
⋮	⋮	⋮

This book should be thought of as an advertisement for other, more advanced texts (or, perhaps, texts where details omitted here are carefully worked out!); there is nothing here that cannot be found elsewhere. What I hope is novel is the emphasis on working with a keyboard at hand to try out computations, the choice of topics, and the commingling of algebra, combinatorics, topology, and geometry. There are all sorts of gaps (some even by design!); for example the Nullstellensatz is not proved, nor is Nakayama's lemma; and little is said about smoothness. The most egregious example of this occurs in Chapter 9, which gives a synopsis of algebraic curves. Since the sketch of Riemann–Roch uses residues, a one-hour turbo lecture on complex analysis is included as an appendix. But generally I have tried to resist the temptation to be completely comprehensive, hoping rather to be convincing without bogging down in detail. The two introductory algebraic geometry texts listed above (Cox–Little–O'Shea and Reid) are nice complementary readings. A good way for readers to begin this book is to flip to Appendix A, which gives a warm-up review of algebra concepts and an introduction to basic Macaulay 2 commands.

These notes grew out of a class taught to junior mathematics majors at Harvard in fall of 2000. I thank Harvard for providing a great postdoctoral experience, the N.S.F. for providing funding, and my students for being such a lively, engaged, hardworking and fun group; Richard Stanley was kind enough to cap the course with a guest lecture. I also thank all the folks from whom I've learned over the years – both in print (see above texts!) and in person. Many people were kind enough to provide feedback on drafts of this book: Marcelo Aguiar, Harold Boas, Al Boggess, Jorge Calvo, Renzo Cavalieri, David Cox, Jim Coykendall, John Dalbec, Marvin Decker, Alicia Dickenstein, David Eisenbud, Bahman Engheta, Chris Francisco, Tony Geramita, Leah Gold, Mark Gross, Brian Harbourne, Mel Hochster, Morten Honsen, Graham Leuschke, Paulo Lima-Filho, John Little, Diane Maclagan, Juan Migliore, Rick Miranda, Alyson Reeves, Vic Reiner, Bill Rulla, Sean Sather-Wagstaff, Fumitoshi Sato, Jessica Sidman, Greg Smith, Jason Starr, Peter Stiller, Emil Straube, Alex Suciu, Hugh Thomas, Stefan Tohaneanu, Will Traves, Adam Van Tuyl, Pete Vermeire, Lauren Williams, and Marina Zompatori. To them, many, many thanks. It goes without saying that any blunders are a result of ignoring their advice. Updates to reflect changes to Macaulay 2, corrections, and (eventually) solutions to the problems will be posted at: http://us.cambridge.org/mathematics/

I owe much to Mike Stillman – teacher, mentor, and friend – who introduced me to most of the material here. I hope that the notes convey some of the enthusiasm and joy in mathematics that Mike imparted to me. To acknowledge my debt (and pay back some small portion!), all author royalties from this book go to the Cornell mathematics department graduate teaching excellence fund.

Chapter 1

Basics of Commutative Algebra

Somewhere early in our mathematical career we encountered the equation

$$f(x, y) = y - x^2 = 0,$$

and learned that the set of points in the plane satisfying this equation (the *zero locus* of f) is a parabola.

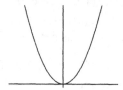

The natural generalization of this problem is to find the solutions to a system of polynomial equations, which is the realm of algebraic geometry. In this chapter we give a whirlwind tour of the basics of commutative algebra. We begin by studying the relationship between an ideal I in a polynomial ring R over a field k, and the set of common zeroes of the polynomials defining I. This object is called a *variety*, and denoted $V(I)$. We prove the Hilbert Basis Theorem, which shows that every ideal in R is finitely generated. Then we tackle the task of breaking a variety into simpler constituent pieces; this leads naturally to the concept of the primary decomposition of an ideal. You may want to warm up by browsing through the algebra appendix if you are hazy on the concepts of group, ring, ideal, and module.

Key concepts: Varieties and ideals, Hilbert Basis Theorem, associated primes and primary decomposition, Nullstellensatz, Zariski topology.

1.1 Ideals and Varieties

Let $R = k[x_1, \ldots, x_n]$ be a polynomial ring over a field k. *Affine n-space* k^n is the set of n-tuples of elements of k. An *affine variety* is the common zero locus of a collection of polynomials $f_i \in R$; the affine variety associated to the set $\{f_1, \ldots, f_m\}$ is written $V(f_1, \ldots, f_m)$. For example, $V(0) = k^n$ and $V(1)$ is the empty set. If you have not done this sort of thing before, try working Exercise A.2.5 in the appendix. Varieties arise quite naturally in many situations. Linear algebra is one special case (the polynomials are all of degree one); other examples of applied problems which involve solving polynomial systems range from computer vision and robot motion to understanding protein placement in cell walls. In fact, this sentence involves varieties: in PostScript, letters are drawn using Bezier cubics, which are parametric plane curves.

Exercise 1.1.1. [23] To define Bezier cubics, we need some terminology. A set $S \subseteq \mathbb{R}^n$ is called *convex* if the line segment between any two points $p, q \in S$ lies in S. Prove that if S is a convex subset of $\mathbb{R}^2$, and $\{p_0, \ldots, p_n\} \subset S$, then any *convex combination* $\sum_{i=0}^{n} t_i \cdot p_i$ with $t_i \geq 0$, $\sum_{i=0}^{n} t_i = 1$ is in S. For four points $p_i = (x_i, y_i)$ in $\mathbb{R}^2$ consider the parametric curve given by:

$$x = x_0(1-t)^3 + 3x_1 t(1-t)^2 + 3x_2 t^2(1-t) + x_3 t^3$$
$$y = y_0(1-t)^3 + 3y_1 t(1-t)^2 + 3y_2 t^2(1-t) + y_3 t^3$$

Prove that p_0 and p_3 lie on the parametric curve, and that the tangent line at p_0 goes through p_1 (chain rule flashback!). Given parametric equations, one might want to find the implicit equations defining an object. These equations can be found by computing a *Gröbner basis*, a technique we'll learn in Chapter 4. ◇

One important observation is that the variety $V(f_1, \ldots, f_m)$ depends only on the ideal I *generated by* $\{f_1, \ldots, f_m\}$. This ideal consists of all linear combinations of $\{f_1, \ldots, f_m\}$ with polynomial coefficients; we write this as $I = \langle f_1, \ldots, f_m \rangle$. The variety $V(f_1, \ldots, f_m)$ depends only on I because if p is a common zero of $f_1, \ldots, f_m$, then p also zeroes out any polynomial combination

$$\sum_{i=1}^{m} g_i(x_1, \ldots, x_n) \cdot f_i(x_1, \ldots, x_n).$$

Thus, we can choose a different set of generators for I without altering $V(I)$. This is analogous to writing a linear transform with respect to different choices of basis. Consider the ideal $I = \langle x^2 - y^2 - 3, 2x^2 + 3y^2 - 11 \rangle$. Take a minute and find $V(I) \subseteq \mathbb{R}^2$. You can do this by just drawing a picture, but you can also do it by renaming x^2 and y^2 and using Gaussian elimination. Of course, this won't work in general. One of our goals will be to find a way to solve such problems systematically, for example, we might want to find a generating set for I where we can read off the solutions. For the ideal above, prove that $I = \langle x^2 - 4, y^2 - 1 \rangle$. This is a set of generators from which it is certainly easy to read off $V(I)$!

Given an ideal J, we have the set of common zeroes $V(J)$, which is a geometric object. Conversely, given $S \subseteq k^n$, we can form the set $I(S)$ of all polynomials vanishing on S. It is easy to check (do so!) that this set is actually an ideal. If $S = V(J)$ for some ideal J, then it is natural to think that $J = I(V(J))$, but this is not the case. For example, if $J = \langle x^2 \rangle \subseteq k[x]$, then $I(V(J)) = \langle x \rangle$. If $f \in J$ and $p \in V(J)$ then by definition $f(p) = 0$. Hence $f \in I(V(J))$, so there is a containment $J \subseteq I(V(J))$.

Exercise 1.1.2. Show that the process of passing between geometric and algebraic objects is inclusion reversing:

$$I_1 \subseteq I_2 \Rightarrow V(I_2) \subseteq V(I_1),$$

and

$$S_1 \subseteq S_2 \Rightarrow I(S_2) \subseteq I(S_1).$$

Use the set $S = \cup\{(0, i)|i \in \mathbb{Z}\} \subseteq \mathbb{R}^2$ to show that it can happen that $S_1 \subsetneq S_2$ but $I(S_1) = I(S_2)$. ◇

For a ring element f and ideal I, a natural algebraic question is: "is $f \in I$?". If we can answer this question on *ideal membership*, then the exercise above shows that there is a geometric consequence: $V(I) \subseteq V(f)$, and we can restrict our search for points of $V(I)$ to points on $V(f)$. So one way to begin to get a handle on a variety is to understand the hypersurfaces on which it sits. Another natural thing to do is to try to break $V(I)$ up into a bunch of more manageable parts. What does "manageable" mean? Well, here is a first candidate:

Definition 1.1.3. *A nonempty variety V is irreducible if it is not the union of two proper subvarieties: $V \neq V_1 \cup V_2$ for any varieties V_i with $V_i \subsetneq V$.*

Theorem 1.1.4. *$I(V)$ is prime iff V is irreducible.*

Proof. First, we need to observe that if X is a *variety*, say $X = V(J)$, then $V(I(X)) = X$. As Exercise 1.1.2 shows, this need not be the case if we only assume X is some set. The inclusion $X \subseteq V(I(X))$ is obvious. By construction $J \subseteq I(X)$, so again by Exercise 1.1.2, $V(I(X)) \subseteq V(J) = X$. We're now ready to prove the theorem. Suppose $I(V)$ is prime but V is reducible with $V = V_1 \cup V_2$. Let $I_1 = I(V_1)$ and $I_2 = I(V_2)$. So there is a point $p \in V_2$ and $f \in I_1$ with $f(p) \neq 0$ (if every $f \in I_1$ vanishes on every $p \in V_2$, then $I_1 \subseteq I_2$, and we'd have a contradiction). By symmetry, there is a $g \in I_2$ and $q \in V_1$ with $g(q) \neq 0$. Clearly $fg \in I(V)$, with neither f nor g in $I(V)$, contradiction. We leave the other direction for the reader. $\square$

As a last warm up before plunging into some proofs, we ask what happens geometrically when we perform standard operations on ideals.

Exercise 1.1.5. Recall that if I and J are ideals, then the sum $I + J = \{f + g \mid f \in I, g \in J\}$ is an ideal, as are $I \cdot J = \langle f \cdot g \mid f \in I, g \in J \rangle$ and $I \cap J$. Show that

$$V(I + J) = V(I) \cap V(J),$$

and that

$$V(I \cdot J) = V(I \cap J) = V(I) \cup V(J). \quad \diamond$$

1.2 Noetherian Rings and the Hilbert Basis Theorem

In the previous section we asked if it was possible to find a "nice" generating set for an ideal. For example, since $k[x]$ is a principal ideal domain, every ideal $I \subseteq k[x]$ has a single generator, which we can find by repeated use of the Euclidean algorithm. So the question of ideal membership is easily solved: once we have a generator for I, to see if $g \in I = \langle h \rangle$, we need only check that h divides g. If we work in rings where ideals can have minimal generating sets which are infinite, then finding a "nice" generating set or running a division algorithm is problematic, so we should begin by finding a sensible class of rings. In this book, *ring* always means *commutative ring with unit*.

Definition 1.2.1. *A ring is Noetherian if it contains no infinite ascending chains (infinite proper inclusions) of ideals, i.e. no sequences of the form*

$$I_1 \subsetneq I_2 \subsetneq I_3 \subsetneq \cdots$$

A module is Noetherian if it contains no infinite ascending chains of submodules. Although this definition seems a bit abstract, it is in fact *exactly* the right thing to make all ideals finitely generated.

Lemma 1.2.2. *A ring is Noetherian iff every ideal is finitely generated.*

Proof. First, suppose every ideal is finitely generated, but that there exists an infinite ascending chain of ideals:

$$I_1 \subsetneq I_2 \subsetneq I_3 \subsetneq \cdots$$

But (check!) $J = \bigcup_{i=1}^{\infty} I_i$ is an ideal. By assumption, J is finitely generated, say by $\{f_1, \ldots, f_k\}$, and each $f_i \in I_{l_i}$ for some l_i. So if $m = max\{l_i\}$ is the largest index, we have $I_{m-1} \subsetneq I_m = I_{m+1} = \cdots$, contradiction. Now suppose that I cannot be finitely generated. By taking a sequence of generators $\{f_1, f_2, \ldots\}$ for I with $f_i \notin \langle f_1, f_2, \ldots f_{i-1} \rangle$, we obtain

$$\langle f_1 \rangle \subsetneq \langle f_1, f_2 \rangle \subsetneq \langle f_1, f_2, f_3 \rangle \subsetneq \cdots,$$

which is an infinite ascending chain of ideals. $\square$

Exercise 1.2.3. Let M be a module. Prove the following are equivalent:

1. M contains no infinite ascending chains of submodules.
2. Every submodule of M is finitely generated.
3. Every nonempty subset Σ of submodules of M has a maximal element (Σ is a partially ordered set under inclusion).

This gives three equivalent conditions for a module to be Noetherian. $\diamond$

Theorem 1.2.4 (Hilbert Basis Theorem). *If A is a Noetherian ring, then so is $A[x]$.*

Proof. Let I be an ideal in $A[x]$. By Lemma 1.2.2 we have to show that I is finitely generated. The set of lead coefficients of polynomials in I generates an ideal I' of A, which is finitely generated (A is Noetherian), say by $g_1, \ldots, g_k$. Now, for each g_i there is a polynomial

$$f_i \in I, f_i = g_i x^{m_i} + \text{terms of lower degree in } x.$$

Let $m = max\{m_i\}$, and let I'' be the ideal generated by the f_i. Given any $f \in I$, we can chop it down by the elements of I'' until its lead term has degree less than m. Consider the A-*module* M generated by $\{1, x, \ldots, x^{m-1}\}$. It is finitely generated, hence Noetherian. So the submodule $M \cap I$ is also

Noetherian. Take generators $h_1, \ldots, h_j$, toss them in with the generators of I'', and we're done. □

Exercise 1.2.5. Prove that if A is Noetherian and M is a finitely generated A-module, then M is Noetherian. Hint: for some n, A^n surjects onto M. What would an infinite ascending chain of submodules of M imply? ◇

In a Noetherian ring, no matter how complicated an ideal I appears to be, there will always be a finite generating set for I. A field k is Noetherian, so the Hilbert Basis Theorem and induction tell us that the ring $k[x_1, \ldots, x_n]$ is Noetherian (of course, so is a polynomial ring over $\mathbb{Z}$ or any other principal ideal domain). Thus, our goal of finding a nice generating set for an ideal does make sense.

1.3 Associated Primes and Primary Decomposition

Throughout this book, we will dwell on the following theme: "To understand a complicated object, break it up into simpler objects". In this section we'll see how to write an ideal in a Noetherian ring in terms of "nice" ideals.

Exercise 1.3.1. (Decomposition I)

1. Prove that $\langle x^2 - 4, y^2 - 1 \rangle$ can be written as the intersection of four *maximal* ideals in $\mathbb{R}[x, y]$. (Hint: what is the corresponding variety?)
2. Prove that $\langle x^2 - x, xy \rangle = \langle x \rangle \cap \langle x - 1, y \rangle$, hence is the intersection of a prime ideal and a maximal ideal in $\mathbb{R}[x, y]$. ◇

The two ideals in Exercise 1.3.1 are intersections of prime ideals (by Exercise A.2.6, maximal ideals are prime). By Theorem 1.1.4 we know that if X is an irreducible variety then $I(X)$ is prime. Since any variety can be written as a union of irreducible varieties, it seems natural to hope that any ideal is an intersection of prime ideals. As $\langle x^2 \rangle \subseteq k[x]$ shows, this hope is vain. However, in a Noetherian ring, any ideal can be written as a finite intersection of irreducible ideals (an *irreducible decomposition*) or as a finite intersection of primary ideals (a *primary decomposition*). Warning: don't confuse an irreducible ideal with an irreducible variety. In fact, it might be good to review the definitions of irreducible and primary ideal at this point (Exercise A.2.5).

Lemma 1.3.2. *In a Noetherian ring R, any ideal is a finite intersection of irreducible ideals.*

Proof. Consider the set Σ consisting of ideals which may not be written as a finite intersection of irreducibles. Since R is Noetherian, Σ has a maximal element I'. But I' is reducible, so we can write $I' = I_1 \cap I_2$, and by assumption I_1 and I_2 are finite intersections (since they properly contain I', and I' is a maximal element of Σ), a contradiction. $\square$

Lemma 1.3.3. *In a Noetherian ring R, irreducible ideals are primary.*

Proof. Let I be irreducible, and suppose $fg \in I$, with $f \notin I$. By passing to the quotient ring $A = R/I$, we only need to show that $g^m = 0$, for some m. There is a chain of ideals in A:

$$0 \subseteq ann(g) \subseteq ann(g^2) \subseteq \cdots,$$

where

$$ann(h) = \{e \in A \,|\, eh = 0\}.$$

Because A is Noetherian, there exists an n such that

$$ann(g^n) = ann(g^{n+1}).$$

Since the zero ideal is irreducible in A and $f \neq 0$, if we can show that $\langle g^n \rangle \cap \langle f \rangle = 0$, we'll be done. So suppose $a \in \langle f \rangle \cap \langle g^n \rangle$; $a \in \langle f \rangle$ implies $ag = 0$. But

$$a \in \langle g^n \rangle \Rightarrow a = bg^n \Rightarrow bg^{n+1} = 0 \Rightarrow bg^n = 0 \Rightarrow a = 0,$$

so indeed $\langle g^n \rangle \cap \langle f \rangle = 0$. $\square$

Primary decompositions are generally used more often than irreducible decompositions, in fact, some books ignore irreducible decompositions completely. The treatment here follows that of [3]; it seems reasonable to include the irreducible decomposition since the proof is so easy! It turns out that primary ideals are very closely related to prime ideals. First, we need a definition:

Definition 1.3.4. *The radical of an ideal I (denoted $\sqrt{I}$) is the set of all f such that $f^n \in I$ for some $n \in \mathbb{N}$; I is radical if $I = \sqrt{I}$.*

Exercise 1.3.5. Prove that if Q is primary, then $\sqrt{Q} = P$ is a prime ideal, and P is the smallest prime ideal containing Q. We say that Q is P-*primary*. Show that if Q_1 and Q_2 are P-primary, so is $Q_1 \cap Q_2$. This is one reason for preferring primary decomposition to irreducible decomposition: the intersection of two irreducible ideals is obviously not irreducible. For the ideal $I = \langle x^2, xy \rangle$, show $\sqrt{I} = \langle x \rangle$ but I is not primary. $\diamond$

A primary decomposition $I = \bigcap_{i=1}^{n} Q_i$ is *irredundant* if for each $j \in \{1, \ldots, n\}$

$$\bigcap_{i \neq j} Q_i \neq I$$

(there are no "extraneous" factors). By Exercise 1.3.5, we may assume that the radicals P_i of the Q_i are distinct; the P_i are called the *associated primes* of I. An associated prime P_i which does not properly contain any other associated prime P_j is called a *minimal* associated prime. The non-minimal associated primes are called *embedded* associated primes. The reason for this terminology is explained in the following example.

Example 1.3.6. Consider the two ideals

$$I_1 = \langle x^2, xy \rangle \text{ and } I_2 = \langle x^2 - x, xy \rangle.$$

Clearly $I_1 = \langle x^2, y \rangle \cap \langle x \rangle$, and $\langle x \rangle$, $\langle x^2, y \rangle$ are primary ideals. So I_1 has one minimal associated prime $\langle x \rangle$ and one embedded associated prime $\langle x, y \rangle$. By Exercise 1.1.5, $V(I \cap J) = V(I) \cup V(J)$. Thus,

$$V(I_1) = V(x) \cup V(x^2, y) = V(x) \cup V(x, y).$$

In the plane, $V(x, y)$ corresponds to the origin, which is "embedded in" the line $V(x)$. Notice that we can write

$$\langle x \rangle \cap \langle x^2, xy, y^2 \rangle = I_1 = \langle x^2, y \rangle \cap \langle x \rangle.$$

Verify that $\langle x^2, xy, y^2 \rangle$ is a primary ideal. This shows that the Q_i which appear in a primary decomposition are not unique. Let's ask the computer algebra package Macaulay 2 to check our work. Appendix A.3 describes how to get started with Macaulay 2; you should glance over the appendix (and, better still, try running the commands) before proceeding.

```
i1 : R=QQ[x,y]

o1 = R

o1 : PolynomialRing

i2 : intersect(ideal(x),ideal(x^2,x*y,y^2))
```

```
          2
o2 = ideal (x*y, x )

o2 : Ideal of R

i3 : intersect(ideal(x),ideal(x^2,y))

          2
o3 = ideal (x*y, x )

o3 : Ideal of R

i4 : o2==o3

o4 = true
```

In Macaulay 2, the command == tests for equality (of course, in this example we could see that the two ideals are equal, but sometimes it won't be so obvious). In Exercise 1.3.12 you'll prove that passing from I to $\sqrt{I}$ causes embedded components to disappear.

```
i5 : radical o2

o5 = ideal x
```

For the ideal I_2 we obtain a primary decomposition

$$I_2 = \langle x \rangle \cap \langle x - 1, y \rangle,$$

hence I_2 has two minimal associated prime ideals, and the primary components are actually prime already, so $\sqrt{I_2} = I_2$.

```
i6 : primaryDecomposition ideal(x^2-x,x*y)

o6 = {ideal (y, x - 1), ideal x}

o6 : List

i7 : (radical ideal(x^2-x,x*y))==ideal(x^2-x,x*y)

o7 = true
```

The zero loci of *all* the primary components of I_1 and I_2 are shown below; the pictures hint that while varieties capture all the geometry of the minimal primes, they forget about embedded primes. Understanding the entire set of primary components of an ideal is part of the motivation for studying *schemes* [34].

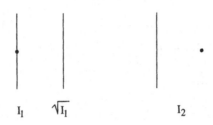

$$I_1 \qquad \sqrt{I_1} \qquad\qquad\qquad I_2$$

Why bother worrying about the embedded primes? Well, for one thing, they carry important information about I. In Chapter 4, we'll learn how to define an order on monomials in a polynomial ring, so that we can define the lead monomial of a polynomial. The set $in(I)$ of all lead monomials of elements of I generates an ideal, and will often have embedded primes *even if I does not*. So what? Well, the point is that many numerical invariants are the same for I and for $in(I)$, but $in(I)$ is often much easier to compute. Punchline: embedded primes matter.

Next we consider how to actually find associated primes and a primary decomposition. A key tool is the operation of *ideal quotient*:

Definition 1.3.7. *Let R be a ring and I, J ideals of R. Then the ideal quotient* $I : J = \{f \in R | f \cdot J \subseteq I\}$.

As usual, you should take a minute and scrawl down a proof that $I : J$ is an ideal (it really will fit in the margin!).

Lemma 1.3.8. *If Q is a P-primary ideal, and $f \in R$, then*

$$f \in Q \Rightarrow \quad Q : f = R$$
$$f \notin Q \Rightarrow Q : f \text{ is } P\text{-primary}$$
$$f \notin P \Rightarrow \quad Q : f = Q$$

Proof. The first statement is automatic, and for the second, if $fg \in Q$, then since $f \notin Q$ we must have $g^n \in Q$ so $g \in P$;

$$Q \subseteq (Q : f) \subseteq P, \text{ so } \sqrt{Q : f} = P,$$

and it is straightforward to show $Q : f$ is P-primary. For the last statement, if $fg \in Q$, then $f^n \notin Q$ (else $f \in P$) so $g \in Q$ and $Q : f \subseteq Q$. $\square$

Exercise 1.3.9. (Distributivity).

1. Show that if a prime ideal $P = P_1 \cap P_2$, then P is one of the P_i.
2. Show that $(I_1 \cap I_2) : f = (I_1 : f) \cap (I_2 : f)$.
3. Show that $\sqrt{I_1 \cap I_2} = \sqrt{I_1} \cap \sqrt{I_2}$. $\diamond$

Lemma 1.3.8 and Exercise 1.3.9 show that in a Noetherian ring, the associated primes of an ideal are independent of the decomposition – in other words, even though the Q_i are not unique, the P_i are! To see this, write

$$I = \bigcap_{i=1}^{n} Q_i,$$

which we can assume is irredundant by the remarks following Exercise 1.3.5. Now, since the decomposition is irredundant, for any j we can find $f_j \notin Q_j$ but which is in all the other Q_i, $i \neq j$. By Lemma 1.3.8 and Exercise 1.3.9, $I : f_j = Q_j : f_j$ is P_j-primary. In particular $\sqrt{Q_j : f_j} = P_j$, which proves:

Lemma 1.3.10. *The associated primes of I are contained in the set $\{\sqrt{I : f} \mid f \in R\}$.*

On the other hand, if P is a prime in the set $\{\sqrt{I : f} \mid f \in R\}$, then it must be associated to I (hint: Exercise 1.3.9).

We can also define the associated primes of a module M. In this case, the set of associated primes $Ass(M)$ consists of primes P such that P is the annihilator of some $m \in M$.

Exercise 1.3.11. ([28], Proposition 3.4) Let M be an R-module, and $S = \{I \subseteq R \mid I = ann(m),$ some $m \in M\}$. Prove that a maximal element of S is prime. $\diamond$

By the previous exercise, the union of the associated primes of M consists precisely of the set of all zero divisors on M. One caution – the associated primes of the module R/I are usually referred to as the associated primes of the ideal I. This seems confusing at first, but is reasonable in the following context: if R is a domain, then no nonzero element of R has nontrivial annihilator. In particular, if $I \subseteq R$ a domain, then *as a module I has no* interesting associated primes. For example, let $R = k[x, y]$, and consider the R-module $M = R/I_1$ with I_1 as in Example 1.3.6. The annihilator of $x \in M$ is $\langle x, y \rangle$, and the annihilator of $y \in M$ is $\langle x \rangle$, so $\{\langle x \rangle, \langle x, y \rangle\} \subseteq Ass(M)$. Is this everything?

Exercise 1.3.12. (Decomposition II).

1. Prove that $\sqrt{I}$ is the intersection of the minimal primes of I.
2. Find (by hand) a primary decomposition for the radical of $\langle y^2 + yz, x^2 - xz, x^2 - z^2 \rangle$.
3. Find a primary decomposition for $\langle xz - y^2, xw - yz \rangle$ as follows: First, observe that when x and y both vanish then both generators of the ideal vanish, so $\langle xz - y^2, xw - yz \rangle \subseteq \langle x, y \rangle$. Use ideal quotient to strip off $\langle x, y \rangle$. You should find that $\langle xz - y^2, xw - yz \rangle : \langle x, y \rangle = \langle xz - y^2, xw - yz, z^2 - yw \rangle$. It turns out (Deus ex machina!) that $J = \langle xz - y^2, xw - yz, z^2 - yw \rangle$ is the kernel of the map

$$R = k[x, y, z, w] \longrightarrow k[s^3, s^2t, st^2, t^3]$$

given by

$$x \to s^3, y \to s^2t, z \to st^2, w \to t^3.$$

Since $R/J \simeq k[s^3, s^2t, st^2, t^3] \subseteq k[s, t]$ and a subring of a domain is a domain, we see that J is a prime ideal, and we have found a primary decomposition $\langle xz - y^2, xw - yz \rangle = J \cap \langle x, y \rangle$. $\diamond$

1.4 The Nullstellensatz and Zariski Topology

Varieties are geometric objects. Given two geometric objects X and Y, it is very natural to ask if there is a map $f : X \to Y$. In analysis we might stipulate that f be continuous or differentiable; the notion of continuity depends on having a *topology*. When X and Y are varieties, one reasonable class of maps to consider are maps which are polynomial (or at least "locally" polynomial). It turns out that there is a specific topology which gives us the right language to study these maps. First, some terminology:

Definition 1.4.1 (Topology). *A topology on a set X is a collection $\mathcal{U}$ of subsets of X which satisfy:*

1. *$\emptyset$ and X are in $\mathcal{U}$.*
2. *$\mathcal{U}$ is closed under finite intersection.*
3. *$\mathcal{U}$ is closed under arbitrary union.*

Members of $\mathcal{U}$ are called the *open sets* of the topology. There is an equivalent formulation using closed sets – a finite union of closed sets is closed, as is any intersection of closed sets. By Exercise 1.1.5, a finite union of affine varieties is itself an affine variety, as is any intersection of affine varieties. This shows that we can define a topology on k^n in which the closed sets are affine varieties. This topology is called the *Zariski topology*, and for this reason the terms affine variety and *Zariski closed* set are used interchangeably. If X is a variety in k^n, then X is endowed with the *subspace topology* – an open set in X is the intersection of X with an open set in k^n. Even though we may not always say it, we'll always have in mind the case where k is algebraically closed (despite the fact that the computations we make are over $\mathbb{Q}$ or a finite field). In this book, when you see $\mathbb{A}_k^n$ think "k^n with Zariski topology", and when you see the word "point", think of a point in the usual topology. If $U \subseteq k^n$ is the complement of the vanishing locus of a polynomial f, then U is called a *distinguished open set*, and written U_f.

Exercise 1.4.2. Show that the distinguished open sets U_f are a *basis* for the Zariski topology on $\mathbb{A}_k^n$: every Zariski open set can be written as a union of distinguished open sets. ◇

The Zariski topology is *quasicompact*: any cover of $\mathbb{A}_k^n$ has a finite subcover. To see this, let $\{U_i\}_{i \in S}$ be a cover of $\mathbb{A}_k^n$ which does not admit a finite subcover. The previous exercise shows that we may suppose the U_i are of the form U_{f_i}. By assumption we can find an infinite sequence $U_{f_1} \subsetneq (U_{f_1} \cup U_{f_2}) \subsetneq \cdots$. Then taking complements of these sets yields an infinite descending chain of varieties $V(f_1) \supsetneq V(f_1, f_2) \supsetneq \cdots$, which is impossible since $k[x_1, \ldots, x_n]$ is Noetherian. A similar argument shows that any subvariety of $\mathbb{A}_k^n$ is quasicompact.

Polynomial functions on k^n obviously restrict to give polynomial functions on a variety $X \subseteq k^n$, and any two polynomials which differ by an element of $I(X)$ define the same function on X. So polynomial functions on an affine variety X correspond to elements of the *coordinate ring* $R/I(X)$. It will be useful to have a local description for this; the reason is that later in the book we shall be constructing objects by patching together *Zariski open* subsets of affine varieties.

Definition 1.4.3. *Let U be an open subset of an affine variety $X \subseteq \mathbb{A}_k^n$, k algebraically closed. A function f is* regular *at a point $p \in U$ if there is a Zariski open neighborhood V of p in X such that $f = \frac{g}{h}$ on V, with $g, h \in k[x_1, \ldots, x_n]/I(X)$, and $h(p) \neq 0$. A function is* regular *on an open set U if it is regular at every point of U.*

A *regular* map is a map defined by regular functions. Two affine varieties X and Y are isomorphic if there exist regular maps $i : X \longrightarrow Y$ and $j : Y \longrightarrow X$ which compose to give the identity.

Exercise 1.4.4. Prove that affine varieties X and Y are isomorphic iff their coordinate rings are isomorphic. (Hint: section 5.4 of [23]). $\diamond$

We'll see shortly that if k is algebraically closed, then the *ring* of regular functions on a distinguished open subset U_f of an affine variety X is isomorphic to $k[x_1, \ldots , x_n, y]/\langle I(X), yf - 1 \rangle$. To prove this, we need to make a detour back to algebra and understand better the relation between J and $I(V(J))$. In §1, we found that $J \subseteq I(V(J))$, and saw that this containment could be proper. From the definition of the radical, $\sqrt{J} \subseteq I(V(J))$. The precise relation between J and $I(V(J))$ follows by first answering the following innocuous question:

When is the variety of an ideal empty?

It is clear that if $1 \in I$ then $V(I)$ is empty, but notice that over a field which is not algebraically closed, $V(I)$ can be empty even if I is a proper ideal (e.g. $\langle x^2 + 1 \rangle \subseteq \mathbb{R}[x]$). However, there is a second beautiful theorem of Hilbert:

Theorem 1.4.5 (Weak Nullstellensatz). *If k is algebraically closed and $V(I)$ is empty, then $1 \in I$.*

To prove the Nullstellensatz properly requires a fair amount of work and is done in almost all books (save this one!) on algebraic geometry; there are nice readable treatments in Chapter 2 of [78] and Chapter 4 of [23], and [28] offers five (!) different proofs. Let's use the Nullstellensatz to answer an earlier question we had:

Theorem 1.4.6 (Strong Nullstellensatz). *If k is algebraically closed and $f \in I(V(I)) \subseteq k[x_1, \ldots, x_n] = R$, then $f^m \in I$, for some m. More tersely put, $\sqrt{I} = I(V(I))$.*

Proof. (The "trick of Rabinowitch"). Given $I = \langle f_1, \ldots, f_j \rangle \subseteq R$ and $f \in I(V(I))$, put $I' = \langle I, 1 - y \cdot f \rangle \subseteq R[y]$. Check that $V(I')$ is empty. So by the weak Nullstellensatz, we can write $1 = \sum a_i \cdot f_i + g(1 - y \cdot f)$. Now just plug in $y = 1/f$ to obtain $1 = \sum a_i(x_1, \ldots, x_n, 1/f) \cdot f_i$, and multiply by a high enough power of f to clean out the denominators. $\square$

With the Nullstellensatz in hand, we can show that if k is algebraically closed, then the ring of regular functions on a distinguished open subset $X_f = U_f \cap X$ of an *irreducible* affine variety $X \subseteq \mathbb{A}_k^n$ is isomorphic to $k[x_1, \ldots, x_n, y] / \langle I(X), yf - 1 \rangle$. Let g be a regular function on X_f. By definition, for each point $p \in X_f$ there is a Zariski open neighborhood U_p of p with $g = \frac{h_p}{k_p}$ on U_p, with h_p and k_p in $R/I(X)$ and k_p nonzero at p. By Exercise 1.4.2 and quasicompactness, we can assume that the cover of X_f is actually finite and given by distinguished open sets $X_{f_i} = X \cap U_{f_i}, i = 1 \ldots j$ with $g = \frac{h_i}{k_i}$ on X_{f_i}. The k_i cannot simultaneously vanish at any point $p \in X_f$, since p lies in some X_{f_m}, and $k_m \neq 0$ on X_{f_m}. So $V(k_1, \ldots, k_j) \cap X_f$ is empty, hence $V(k_1, \ldots, k_j) \cap X \subseteq V(f)$. By the Nullstellensatz, there exist l_i with $f^m = \sum_{i=1}^{j} l_i k_i$ (the equations defining $I(X)$ are implicit in this expression, because the k_i are defined modulo $I(X)$). Since $\frac{h_i}{k_i} = \frac{h_j}{k_j}$ on $X_{f_i} \cap X_{f_j}$, on the common intersection of *all* the X_{f_i} we can write

$$f^m \cdot g = \sum_{i=1}^{j} l_i k_i \frac{h_i}{k_i}.$$

By Lemma 1.3.8 and Lemma 1.4.7 (below), the common intersection of the X_{f_i} is Zariski dense (we assumed X irreducible). Thus, the expression above is actually valid on all of X_f, so we can write g as an element of $R/I(X)$ over f^m, as claimed. Setting $f = 1$ shows that the ring of functions regular everywhere on a variety $X \subseteq \mathbb{A}_k^n$ is simply $R/I(X)$. The hypothesis that X is irreducible can be removed, but the proof is a bit more difficult: see [53], II.2.2.

For any set $S \subseteq \mathbb{A}_k^n$, Exercise 1.1.2 shows that $V(I(S))$ is the smallest variety containing S. So in the Zariski topology $V(I(S))$ is the *closure* of S; we write $\overline{S}$ for $V(I(S))$ and call $\overline{S}$ the *Zariski closure* of S. For $S \subseteq \mathbb{R}^2$ as in Exercise 1.1.2, $\overline{S} = V(x)$. A second nice application of the Nullstellensatz relates the Zariski closure of a set and the ideal quotient. Lemma 1.3.8 tells us that ideal quotient can be used to pull apart the irreducible pieces of an ideal. As an example, compute $\langle xy \rangle : \langle x \rangle$ and $\langle x^2, xy \rangle : \langle x \rangle$. What you should see is the following:

$$\langle xy \rangle \quad : \quad \langle x \rangle \quad = \quad \langle y \rangle \qquad\qquad \langle x^2, xy \rangle : \langle x \rangle \quad = \langle x, y \rangle$$

The picture on the left makes perfect sense, but the picture on the right is meant to make you think. How does it relate to primary decomposition?

Lemma 1.4.7.

$$\overline{V(I) - V(J)} \subseteq V(I : J),$$

and if k is algebraically closed and I is radical, then this is an equality.

Proof. By Exercise 1.1.2, we need to show $I : J \subseteq I(V(I) - V(J))$. So let $f \in I : J$, and take $p \in V(I) - V(J)$. Since $p \notin V(J)$, there is a $g \in J$ with $g(p) \neq 0$. From the definition of ideal quotient, $f \cdot g$ is in I, and so $p \in V(I)$ means $f(p) \cdot g(p) = 0$, and we're over a field, so this shows that $\overline{V(I) - V(J)} \subseteq V(I : J)$. For the second part, since k must be algebraically closed, you can guess that the Nullstellensatz plays a role. Figure it out!   □

Example 1.4.8. Let $S = \{p_1, \ldots, p_4\} = \{(0, 0), (0, 1), (1, 0), (1, 1)\} \subseteq \mathbb{A}_k^2$ be a set of four points in the affine plane. Then

$$I(S) = \bigcap_{i=1}^{4} I(p_i) = \langle x^2 - x, y^2 - y \rangle.$$

To remove the points lying on the line $V(x - y)$, we need to form $I(S) :$ $\langle x - y \rangle$, the result should be the ideal of the two remaining points.

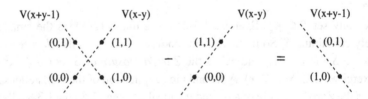

```
i8 : ideal(x^2-x,y^2-y):ideal(x-y)

                    2
o8 = ideal (x + y - 1, y  - y)

o8 : Ideal of R
```

We've been computing radicals, intersections, quotients, and primary decompositions using Macaulay 2, with no discussion of the underlying algorithms. Chapter 4 gives an overview of Gröbner basis techniques, which is the engine behind the computations. For a comprehensive treatment we recommend [23].

This chapter covers the bare essentials of commutative algebra. It is not a substitute for a course in commutative algebra, but rather attempts to hit the high points we'll need in the rest of the book. Good additional sources are Atiyah–Macdonald [3] chapters 1,4,6,7, Cox–Little–O'Shea [23] chapters 1 and 4, Eisenbud [28] chapters 0,1,3, and Smith–Kahanpää–Kekäläinen–Traves [86] chapters 1,2,4. To learn more about the Zariski topology and regular functions, see [28], Exercise 1.24, Chapter 2 of [52], or Chapter 4 of [86].

Chapter 2

Projective Space and Graded Objects

If $f(x)$ is a polynomial with real coefficients, $f(x)$ may have no real roots. We remedy this by passing to the algebraic closure $\mathbb{C}$; since $\mathbb{R} \subseteq \mathbb{C}$ we don't lose any information in doing so. A similar analogy can be used to motivate the construction of *projective space*, which is a natural compactification of affine space. If f and g are elements of $\mathbb{C}[x, y]$, $V(f, g) \subseteq \mathbb{A}^2_{\mathbb{C}}$ may be empty. For example, this is the case if $V(f)$ and $V(g)$ are two parallel lines. On the other hand, in the projective plane $\mathbb{P}^2_{\mathbb{C}}$, not only is $V(f, g)$ nonempty, it actually consists of *exactly* the right number of points. We'll make all this precise in a bit, but the idea is that from a geometric perspective, projective space is often the right place to work.

In order to make sense of varieties in projective space, we have to study homogeneous polynomials, so we introduce the concept of graded rings and modules. Just as beautiful geometric theorems hold in projective space, beautiful algebraic theorems hold for graded rings and modules, highlighting the interplay between algebra and geometry. We define the Hilbert function and Hilbert polynomial; a key tool in computing these objects is the notion of an exact sequence, so we also take some first steps in homological algebra.

Key concepts: Projective space, graded module, chain complex, homology, exact sequence, Hilbert function, Hilbert polynomial, Hilbert series.

2.1 Projective Space and Projective Varieties

Over an algebraically closed field (which is primarily what we'll have in mind throughout this book) n-dimensional affine space $\mathbb{A}^n_k$ can be thought of as plain old k^n. Projective n-dimensional space (denoted $\mathbb{P}^n_k$) is just $\mathbb{A}^{n+1}_k$ minus the origin, modulo the relation

$$(a_0, \ldots, a_n) \sim (b_0, \ldots, b_n) \Leftrightarrow (a_0, \ldots, a_n) = \lambda \cdot (b_0, \ldots, b_n), \lambda \in k^*.$$

In English, the relation simply says that we are identifying any two points which lie on the same line through the origin. A point of $\mathbb{P}^n_k$ has *homogeneous coordinates* $(a_0 : \ldots : a_n)$ defined up to nonzero scalar, in particular, points in $\mathbb{P}^n_k$ are in one to one correspondence with lines through the origin in $\mathbb{A}^{n+1}_k$. A very useful way to think of $\mathbb{P}^n_k$ is as

$$\mathbb{A}^n_k \cup \mathbb{P}^{n-1}_k.$$

To see this, take a line $(a_0 : \ldots : a_n) := \lambda(a_0, \ldots, a_n)$. If $a_0 \neq 0$, then scale a_0 to one and use $(a_1/a_0, \ldots, a_n/a_0)$ as coordinates. The condition that $a_0 \neq 0$ means that we are on the complement of $V(x_0)$, which is a Zariski open set of $\mathbb{P}^n_k$ (see 2.1.3), isomorphic to $\mathbb{A}^n_k$. The coordinates are exactly those above; once we scale the first coordinate to one, the point is no longer free to move. If $a_0 = 0$, then we can forget it, and we're on the Zariski closed set $V(x_0)$, where a typical element may be written as $(0 : b_1 : \ldots : b_n)$. Of course, scaling can never change the first coordinate of the corresponding line to a nonzero value, so $V(x_0)$ corresponds to $\mathbb{P}^{n-1}_k$. We can visualize the projective plane as

$$\mathbb{P}^2_k = \mathbb{A}^2_k \cup \mathbb{P}^1_k.$$

The aesthete has already encountered $\mathbb{P}^2_k$ in a first course in art history – a nonmathematical motivation for the construction of the projective plane is simply that it is how the world appears to us. Italian renaissance artists used perspective to make their paintings more lifelike (perhaps foreshadowing Italian virtuosity in algebraic geometry!). The idea is simple: if you stand on a set of railroad tracks in the middle of the Kansas plains, you seem to be standing on a flat plane; the railroad tracks appear to meet at the horizon. A perfectly straight, infinite set of tracks would appear from outer space to be a great circle. If you watch a train recede into the distance on such a set of tracks, then that train will reappear directly behind you. So the vanishing point in front of you and the point on the horizon behind you must be the same point. Since you see the whole horizon by rotating through 360 degrees, the horizon forms a circle, with antipodal points identified. In other words, in the projective plane, two parallel lines meet at the horizon, and two antipodal points on the horizon are identified.

How can we define a variety in projective space? Since every point on a line through the origin in $\mathbb{A}^{n+1}_k$ is identified, if we want a projective variety to be the zero locus of a set of polynomials, we'll need to restrict to polynomials f such that for all $\lambda \in k^*$:

$$f(p) = 0 \Rightarrow f(\lambda \cdot p) = 0.$$

A polynomial f is *homogeneous* if all the monomials appearing in f have the same degree, so $x^3 + xyz + yz^2$ is homogeneous, and $y - x^2$ is not. Take a minute to verify that if k is infinite, then the condition above is satisfied iff f is a homogeneous polynomial. Of course, a homogeneous polynomial $f \in k[x_0, \ldots, x_n]$ describes a variety $V(f)$ in both $\mathbb{P}_k^n$ and $\mathbb{A}_k^{n+1}$. For example, when $n = 2$, $V(f)$ defines a curve in $\mathbb{P}^2$ or a surface in $\mathbb{A}^3$ – take a look at the picture below to see why. At this point, it is reasonable to ask: *why define projective space?* $\mathbb{A}_k^n$ seems natural, while $\mathbb{P}_k^n$ seems contrived. Here is a beautiful theorem that really requires that we work in projective space:

Theorem 2.1.1 (Bezout's theorem). *If k is an algebraically closed field and f, g are homogeneous elements of $k[x, y, z]$ of degrees d and e with no common factor, then the curves in $\mathbb{P}_k^2$ defined by f and g meet in $d \cdot e$ points, counted with multiplicity (ignore "multiplicity" for now–it is explained in Example 2.3.10).*

Example 2.1.2. Example 1.4.8, revisited. Let $V = V(x(x - z), y(y - z))$. If we consider V as a variety in $\mathbb{P}_k^2$, then the projective curves $V(x(x - z))$ and $V(y(y - z))$ intersect in four points, which lie on the affine plane where $z = 1$:

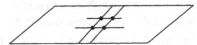

If instead we consider V as a variety in $\mathbb{A}_k^3$, then V consists of four lines through the origin, as below. The dots indicate where the four lines meet the plane $z = 1$, i.e., how the affine picture below relates to the projective picture above:

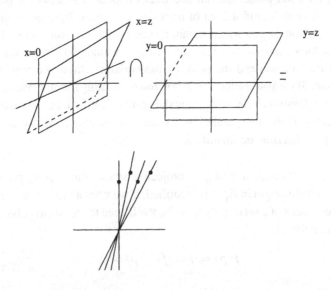

Definition 2.1.3. *A homogeneous ideal is an ideal which can be generated by homogeneous elements. A variety in projective space (or projective variety) is the common zero locus of a homogeneous ideal. The Zariski topology on $\mathbb{P}^n$ is defined by making projective varieties the closed sets.*

We close by mentioning a nice way to visualize real projective space. Since a line through the origin in $\mathbb{A}_{\mathbb{R}}^{n+1}$ hits the unit n-sphere in two points, we can also think of real projective n-space as the unit n-sphere with antipodal points identified. For example, to see why two parallel lines in $\mathbb{P}_{\mathbb{R}}^2$ meet, consider the affine planes $V(x)$ and $V(x - z)$ (leftmost figure above). When we intersect these two planes with the unit sphere, we obtain two great circles:

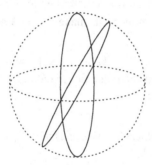

In $\mathbb{A}_{\mathbb{R}}^3$, the common intersection of the two planes and the unit sphere consists of the points $(0, 1, 0)$ and $(0, -1, 0)$; in $\mathbb{P}_{\mathbb{R}}^2$ these two points are identical, and so two parallel lines meet in a single point. Notice that the point lies on the line at infinity $V(z) \subseteq \mathbb{P}_{\mathbb{R}}^2$.

2.2 Graded Rings and Modules, Hilbert Function and Series

When someone mentions the degree of a polynomial, everyone knows what it is. Let's make it formal: A $\mathbb{Z}$-graded ring R is a ring with a direct sum decomposition (as an abelian group) into homogeneous pieces

$$R = \bigoplus_{i \in \mathbb{Z}} R_i,$$

such that if $r_i \in R_i$ and $r_j \in R_j$, then $r_i \cdot r_j \in R_{i+j}$. There are more general definitions; for example, instead of using $\mathbb{Z}$ you can grade by an arbitrary group, but when we talk about graded, we'll assume the group is $\mathbb{Z}$. Of course, if you have a module M over a graded ring R, the module will be graded if

$$M = \bigoplus_{i \in \mathbb{Z}} M_i,$$

and if you multiply an element of R_i and an element of M_j, the result is in M_{i+j}. An element of a graded module is called homogeneous of degree i if it is an element of M_i. The most common examples of graded rings are $R = k[x_1, \ldots, x_n]$ (where k is itself a ring, called the *ring of coefficients*), and any quotient of R by a homogeneous ideal I. In $k[x, y]$, is $xy + y^2$ a homogeneous element? How about $x + 1$? Why is $k[x, y]/\langle y^2 - 1\rangle$ not a graded ring? If $R_0 = k$ is a field (which for us will always be the case unless otherwise noted), then each graded piece of the ring is also a k-vector space, so it has a dimension. For example, the dimension of $k[x]_i$ is one, for all i, and the dimension of $k[x, y]_i$ is $i + 1$. To see this, write out bases for the first few degree pieces; using monomials makes the task easy.

Exercise 2.2.1. Prove that $dim_k(k[x_0, \ldots, x_n]_i) = \binom{n+i}{i}$. If you know about tensor products, see if you can relate this to them. If not, fear not. Tensors are covered in Chapter 6. $\diamond$

Definition 2.2.2. *The Hilbert function of a finitely–generated, graded module M is $HF(M, i) = dim_k M_i$.*

One of the most important examples of a graded module is just the ring itself, but with the grading shifted. Let $R(i)$ denote R (considered as a module over itself), but where we think of the generator as being in degree $-i$. This seems counterintuitive, but the notation works well, because $R(i)_j = R_{i+j}$. For example, $k[x, y](-2)$ "looks like"

$$
\begin{array}{lccccccc}
\text{degree } i & = & 0 & 1 & 2 & 3 & \cdots \\
\text{basis of } k[x, y]_i & = & 1 & x, y & \cdots & \cdots & \cdots \\
\text{basis of } k[x, y](-2)_i & = & 0 & 0 & 1 & x, y & \cdots
\end{array}
$$

Example 2.2.3. Let $R = k[x, y, z]$ and $I = \langle x^3 + y^3 + z^3\rangle$. The dimension of R/I in degree i will be the dimension of R in degree i minus the dimension of I in degree i. I is a principal ideal, generated in degree 3, so the degree i piece of I just looks like the degree $i - 3$ piece of R. Thus:

i	0	1	2	3	4	5	6	$\cdots$
HF$(R/I, i)$	1	3	6	9	12	15	18	$\cdots$

As i gets large (in fact, as soon as $i \geq 1$), the dimension of $(R/I)_i$ is just

$$\dim_k k[x, y, z]_i - \dim_k k[x, y, z]_{i-3} = \binom{i+2}{2} - \binom{i-1}{2} = 3i.$$

Now let's add a linear form to I – to make life easy, say the form is x, and put $J = I + \langle x \rangle$. Since $R/J \simeq k[y, z]/\langle y^3 + z^3 \rangle$, the Hilbert function is:

i	0	1	2	3	4	...
HF($R/J, i$)	1	2	3	3	3	...

By Bezout's theorem, a line and a cubic curve in $\mathbb{P}_k^2$ meet in three points. Do you have a guess about how this relates to the dimension of a high degree piece of the quotient ring? Try your guess on some other examples (pick equations to minimize your work). Another way of encoding the data of the Hilbert function is via a formal power series, called the Hilbert (or Hilbert–Poincaré) series:

Definition 2.2.4. *The Hilbert series of a finitely–generated, graded module M is*

$$HS(M, t) = \sum_{i \in \mathbb{Z}} HF(M, i)t^i.$$

We'll see in the next chapter that if M is a finitely–generated, graded module over $k[x_1, \ldots, x_n]$, then $HS(M, t) = P(t)/(1 - t)^n$ with $P(t) \in \mathbb{Z}[t, t^{-1}]$. In Macaulay 2, the Hilbert function and Hilbert series are easy to compute. We illustrate for the previous example:

```
i1 : R=ZZ/101[x,y,z];

i2 : I = matrix {{x^3+y^3+z^3}}

o2 = {0} | x3+y3+z3 |

                1        1
o2 : Matrix R   <--- R

i3 : hilbertFunction(3,coker I)

o3 = 9
```

The semicolon after the ring declaration prevents Macaulay 2 from echoing, so the following output is suppressed:

```
o1 = R
```

```
o1 : PolynomialRing
```

The `hilbertFunction` command is self-explanatory: it expects as input a degree and a finitely–generated, graded R-module M. If $R = k[x_1, \ldots, x_n]$, then the command `poincare` M returns the numerator of the Hilbert series of M, but in unsimplified form: the denominator of the Hilbert series is understood as $(1 - t)^n$. For example, if $R = k[x, y, z]$, then $HS(R, t) = \frac{1}{(1-t)^3}$ (see next exercise!), so `poincare` R returns 1.

```
i4 : poincare R
```

```
o4 = 1
```

```
o4 : ZZ[ZZ^1]
```

```
i5 : poincare coker I
```

```
             3
o5 = 1 - $T
```

```
o5 : ZZ[ZZ^1]
```

```
i6 : J = matrix {{x,x^3+y^3+z^3}}
```

```
o6 = | x x3+y3+z3 |
```

```
           1       2
o6 : Matrix R   <--- R
```

```
i7 : poincare coker J
```

```
           3     4
o7 = 1 - $T - $T  + $T
```

So

$$HS(R/J, t) = \frac{1 - t - t^3 + t^4}{(1 - t)^3} = \frac{1 + t + t^2}{1 - t}.$$

Exercise 2.2.5. Hilbert series for $k[x_1, \ldots, x_n]$. If $n = 1$, then each graded piece has dimension one, so

$$HS(k[x], t) = 1 + t + t^2 + \ldots = \frac{1}{1-t}.$$

Prove that

$$HS(k[x_1, \ldots, x_n], t) = \frac{1}{(1-t)^n}. \quad \diamond$$

Recall that a ring (module) is Noetherian if there are no infinite proper *ascending* chains of ideals (submodules). What about *descending* chains? A ring (or module) is *Artinian* if there are no infinite proper *descending* chains of ideals (submodules). Now, suppose we have a graded ring R. If it does not die out in high degree, then we can cook up an infinite descending chain of ideals by taking successive graded pieces of the ring:

$$\langle R_1 \rangle \supsetneq \langle R_2 \rangle \supsetneq \langle R_3 \rangle \cdots$$

In particular, if R is a polynomial ring and M a finitely–generated, graded R–module, then M is Artinian iff $M_i = 0$ for i sufficiently large, so such a module is Artinian iff the Hilbert series is actually a polynomial in $\mathbb{N}[t, t^{-1}]$. Consider $R = k[x, y]/\langle x^2, y^2 \rangle$. We have

$$
\begin{array}{lllllll}
\text{degree } i & = & 0 & 1 & 2 & 3 & \ldots \\
\text{basis of } R_i & = & 1 & x, y & xy & 0 & \ldots
\end{array}
$$

Thus, the Hilbert series of R is $1 + 2t + t^2$.

```
i1 :   R=ZZ/31991[x,y];

i2 : poincare coker matrix {{x^2,y^2}}

              2      4
o2 = 1 - 2$T  + $T

i3 : factor o2

             2           2
o3 = (- 1 + $T) (1 + $T)

i4 : Q = R/ideal(x^2,y^2)
```

```
         R
o4 =  --------
        2   2
       (x , y )
```

o4 : QuotientRing

i5 : poincare Q

```
            2      4
o5 = 1  -  2$T   + $T
```

So as expected

$$HS(R/I, t) = \frac{1 - 2t^2 + t^4}{(1-t)^2} = 1 + 2t + t^2.$$

We can compute this by considering R/I as an R-module or as a graded ring in its own right.

Exercise 2.2.6. Compute the Hilbert series of

$$k[x, y, z]/\langle x^2, y^3, z^4 \rangle.$$

Can you see how to compute the Hilbert series of

$$k[x_1, \ldots, x_n]/\langle x_1^2, x_2^3, \ldots, x_n^{n+1} \rangle?$$

A *monomial ideal* is an ideal which can be generated by monomials. If I is monomial, when is R/I Artinian? ◇

2.3 Linear Algebra Flashback, Hilbert Polynomial

In the last section, in Example 2.2.3 we observed that the Hilbert function was actually a polynomial function of i, at least when i was sufficiently large. This is no accident, and the proof is a very nice illustration of the reason that it is important to study *graded* maps of graded modules. A homomorphism of graded modules is called graded if it preserves the grading, i.e. $M \xrightarrow{\phi} N$ is graded if for all i

$$\phi(M_i) \subseteq N_i.$$

The basic motivation is bookkeeping – by requiring maps to be graded, we obtain sequences where it makes sense to look at what is happening in a single degree. But the degree i piece of a graded module is simply a vector space, so we are reduced to linear algebra! This is the raison d'être of graded maps. We begin with a quick linear algebra review. A sequence of vector spaces and linear transforms

$$V: \ \cdots \xrightarrow{\phi_{j+2}} V_{j+1} \xrightarrow{\phi_{j+1}} V_j \xrightarrow{\phi_j} V_{j-1} \xrightarrow{\phi_{j-1}} \cdots$$

is called a *complex* (or chain complex) if

$$\text{image } \phi_{j+1} \subseteq \text{kernel } \phi_j.$$

The sequence is *exact* at position j if image $\phi_{j+1} = $ kernel ϕ_j; a complex which is exact everywhere is called an *exact sequence*. We define the *homology* of the complex V as

$$H_j(V) = \text{kernel } \phi_j / \text{image } \phi_{j+1}.$$

Exercise 2.3.1. Complexes

1. Compute the homology of the complex

$$0 \longrightarrow V_1 \xrightarrow{\phi} V_0 \longrightarrow 0,$$

where $V_1 = V_0 = k^3$ and ϕ is:

$$\begin{bmatrix} 1 & 0 & -1 \\ -1 & 1 & 0 \\ 0 & -1 & 1 \end{bmatrix}$$

2. Show that for a complex $V: 0 \longrightarrow V_n \longrightarrow \cdots \longrightarrow V_0 \longrightarrow 0$ of finite-dimensional vector spaces,

$$\sum_{i=0}^{n} (-1)^i \dim V_i = \sum_{i=0}^{n} (-1)^i \dim H_i(V).$$

The alternating sum above is called the *Euler characteristic* of V, and written $\chi(V)$. So if V is exact then $\chi(V) = 0$. ◇

   The definitions above (complex, homology, exact) all generalize in the obvious way to sequences of modules and homomorphisms; when working with graded modules we require that the maps are also graded.

Example 2.3.2. Let R be a polynomial ring (regarded as a module over itself), and $f \in R_i$. Consider the map:

$$R \xrightarrow{\cdot f} R.$$

This is *not* a map of graded modules, because it sends 1 (a degree zero object) to f (a degree i object). But if we declare 1 to have degree i in our source module, then we do have a graded map:

$$R(-i) \xrightarrow{\cdot f} R.$$

   For emphasis, we say it again: when R is a polynomial ring over a field, then studying graded maps between graded R-modules is nothing more than linear algebra! Now we prove our earlier observation that the Hilbert function becomes a polynomial, for $i \gg 0$.

Theorem 2.3.3. *If M is a finitely generated, graded module, then there exists a polynomial $f(x) \in \mathbb{Q}[x]$ such that for $i \gg 0$, $HF(M, i) = f(i)$. The polynomial $f(i)$ is called the Hilbert polynomial of M, written $HP(M, i)$.*

Proof. Induct on the number of variables in the ring over which M is defined, the base case being trivial. So, suppose it is true in $n - 1$ variables. We can build an exact sequence:

$$0 \longrightarrow K \longrightarrow M(-1) \xrightarrow{\cdot x_n} M \longrightarrow C \longrightarrow 0,$$

where K and C are the kernel and cokernel of the map given by multiplication by x_n. K and C are finitely generated, and since x_n kills both K and C, they are actually finitely generated modules over a polynomial ring in $n - 1$ variables. Thus $HF(M, i) - HF(M, i - 1) \in \mathbb{Q}[i], i \gg 0$. Now do Exercise 2.3.4. $\square$

Exercise 2.3.4. A function $P : \mathbb{N} \to \mathbb{Z}$ such that $\Delta P(i) := P(i) - P(i - 1)$ is a polynomial with rational coefficients (for i sufficiently large) is itself a polynomial with rational coefficients, and has degree one greater than ΔP. Hint: induct on the degree s of the difference polynomial. The base case is trivial. If $\Delta P(i) = a_s i^s + \ldots$, define $h = a_s s! \binom{i}{s+1}$, and compute Δh. By construction, $\Delta P - \Delta h$ will have degree $s - 1$. $\Diamond$

Exercise 2.3.5. (Hilbert polynomial for a set of points).

1. For a single point $p \in \mathbb{P}_k^n$, compute $HP(R/I(p), i)$. Hint: you may as well assume the point is given by $(0:\ldots:0:1)$.

2. If $I(p_i)$ denotes the ideal of a point, prove that the sequence

$$0 \longrightarrow I(p_1) \cap I(p_2) \longrightarrow I(p_1) \oplus I(p_2) \xrightarrow{\phi} I(p_1) + I(p_2) \longrightarrow 0,$$

   is exact, where $\phi(f, g) = f - g$.

3. What is $I(p_1) + I(p_2)$? What is $HP(R/\langle I(p_1) + I(p_2)\rangle, i)$? Use induction to show that for distinct points $\{p_1, \ldots, p_d\} \in \mathbb{P}_k^n$,

$$HP\left(R\Big/\Big\langle \bigcap_{j=1}^{d} I(p_j)\Big\rangle, i\right) = d. \quad \diamond$$

The main reason that the Hilbert polynomial of R/I is important is that it contains all sorts of useful geometric information about $V(I)$. We've seen one simple instance of this above. If $V(I)$ is a projective variety defined by a homogeneous ideal I, then in Chapter 3 we'll see that the Hilbert polynomial of R/I can be written as an alternating sum of binomial coefficients. So there will exist $a_i \in \mathbb{Z}$, $a_m > 0$ such that the Hilbert polynomial will have the form

$$\frac{a_m}{m!} i^m + \frac{a_{m-1}}{(m-1)!} i^{m-1} + \cdots.$$

Definition 2.3.6. *For a homogeneous ideal $I \subseteq k[x_0, \ldots, x_n]$ with*

$$HP(R/I, i) = \frac{a_m}{m!} i^m + \frac{a_{m-1}}{(m-1)!} i^{m-1} + \cdots,$$

we define the dimension of the projective variety $V(I) \subseteq \mathbb{P}_k^n$ as m, the codimension of I as the complementary dimension of $V(I) = n - m$, and the degree of $V(I)$ as a_m.

We have an intuitive notion of dimension – basically, from vector calculus and the implicit function theorem, the dimension of a geometric object is the dimension of the tangent space at a smooth point. This point of view is very important and is nicely explained in both [23] and [78]. But there is another natural way to think of dimension – if we slice an object with a *generic* hyperplane, then the dimension of the slice should be one less than the dimension of the object – if we slice a surface in three space with a generic plane, we get a curve. Now, the dimension should be the number of times you can slice till you end up with a collection of points. The *degree* of the variety is simply the number of points, as counted by the Hilbert polynomial of the resulting zero-dimensional object. This agrees with our notion of the degree of a curve in the projective plane. This definition also works fine for projective space: from Exercise 2.2.1, $\mathbb{P}^n$ has dimension n and degree one. If $HP(R/I, i) = 0$ then R/I is Artinian; to make the definition of codimension work out right we decree the zero polynomial to have degree -1. In the next chapter we will show that the intuitive notion corresponding to slicing down with hyperplanes and the algebraic formulation in terms of the Hilbert polynomial coincide; for now we content ourselves with an example.

Example 2.3.7. Consider the variety in $\mathbb{P}^3$ defined by the ideal

$$\langle w^2 - yw, xw - 3zw, x^2y - y^2z - 9z^2w + zw^2, x^3 - 3x^2z - xyz + 3yz^2 \rangle$$

```
i1 : R=ZZ/101[x,y,z,w];

i2 : m=matrix{{w^2-y*w, x*w-3*z*w,
        x^2*y-y^2*z-9*z^2*w+z*w^2,
        x^3-3*x^2*z-x*y*z+3*y*z^2}}

o2 = | -yw+w2  xw-3zw  x2y-y2z-9z2w+zw2
        x3-3x2z-xyz+3yz2 |

                1        4
o2 : Matrix R   <--- R

i3 : hilbertPolynomial coker m

o3 = - P   + 3*P
        0        1
```

Macaulay 2 gives the Hilbert polynomial in terms of projective spaces: read P_n as $\binom{n+i}{i}$. In this example, $-P_0 + 3 * P_1 = -1 + 3(i + 1) = 3i + 2$.

```
i4 : I=ideal m;

o4 : Ideal of R

i5 : codim I

o5 = 2
```

```
i6 : degree I

o6 = 3

i7 : primaryDecomposition I

                 2
o7 = {ideal (w, x  - y*z), ideal (y - w, x - 3z)}
```

$V(I)$ has two irreducible pieces: the first is a plane conic curve lying in the plane where $w = 0$, and the second is the projective line given by the vanishing of two linear forms. Thus, this should have degree 3 and dimension 1, which is indeed the case. We slice with a generic hyperplane to double check.

```
i8 : lin=ideal random(R^{1},R^1)

o8 = ideal(42x-50y+39z+9w)

o8 : Ideal of R

i9 : slice=I+lin;

o9 : Ideal of R

i10 : hilbertPolynomial coker gens slice

o10 = 3*P
         0
```

A few words are in order here: slice is an ideal, and to compute a quotient (cokernel), Macaulay 2 expects a matrix. The gens command turns the ideal into a matrix of elements, for which coker makes sense. Try the above sequence without gens and see what happens.

Exercise 2.3.8. For $R = k[x, y, z]$ and $I = \langle x^2 - xz, y^3 - yz^2 \rangle$, ask Macaulay 2 to compute the Hilbert polynomial of R/I. Draw a picture of the variety in $\mathbb{P}^2$ (work on the patch where $z = 1$), and verify that Bezout's theorem holds. ◇

The most naive possible generalization of Bezout's theorem is false: suppose $\{f_1, \ldots, f_n\} \subseteq k[x_0, \ldots, x_n]$ are polynomials which have no pairwise common factor; say degree $f_i = d_i$. As the next example illustrates, it is not in general true that

$$V(f_1, \ldots, f_n) \subseteq \mathbb{P}^n$$

is a set of $d_1 \cdot d_2 \cdots d_n$ points.

Example 2.3.9. (The twisted cubic, revisited) In Exercise 1.3.12 we encountered the ideal

$$I = \langle xz - y^2, xw - yz, z^2 - yw \rangle.$$

It is easy to check that these polynomials have no pairwise common factors. If we work on the affine patch U_x where $x = 1$, then the equations are

$$z = y^2, \, w = y^3.$$

So on U_x the zero locus is given parametrically as $(1, y, y^2, y^3)$; in particular the zero locus is one (rather than zero) dimensional. Let's see if we can guess the Hilbert polynomial. On the patch U_x, a generic hyperplane will have the form $a_0 + a_1 y + a_2 z + a_3 w = 0$ (remember, we're in affine space now), and the common zero locus of the hyperplane and the curve is

$$a_0 + a_1 y + a_2 y^2 + a_3 y^3 = 0.$$

As long as the ground field is algebraically closed, we expect the hyperplane to meet the curve in three points, so we expect that

$$HP(R/I, i) = 3i + a$$

for some constant a. Use Macaulay 2 to compute the Hilbert polynomial, and then verify the computation by appealing to the isomorphism of 1.3.12.

Example 2.3.10. (Nonreduced points) In $k[x, y, z]$, the polynomials $y^2 - xz$ and x have no common factor, so Bezout's theorem tells us that $V(y^2 - xz)$ and $V(x)$ should meet in two points *if we count with multiplicity*. What does this mean? Well, since $I = \langle y^2 - xz, x \rangle = \langle y^2, x \rangle$, we see that the Hilbert polynomial of R/I is 2. The only point where the curves actually meet is $(0 : 0 : 1)$; draw a picture on the affine patch where $z = 1$ and you'll see that the curves are tangent at this point. Intuitively, if we perturb the line, it will meet the conic in two distinct points. The key idea is that the Hilbert polynomial "sees" this higher order of contact and counts it correctly; this is what counting with multiplicity means. Here are two more examples of codimension two ideals in $k[x, y, z]$ whose underlying variety is a point in $\mathbb{P}^2$, but where that variety has forgotten interesting structure–more supporting evidence for studying *schemes*!

First, consider the ideal

$$L = \langle y^2, x^2 \rangle \subseteq k[x, y, z].$$

It is clear that $\sqrt{L} = \langle y, x \rangle$. Again, it is easy to compute $HP(R/L, i)$, since for $i \geq 2$ a basis for $(R/L)_i$ is

$$\{z^i, z^{i-1}x, z^{i-1}y, z^{i-2}xy\}.$$

If we think of $V(x^2)$ as two infinitesimally near parallel lines, and similarly for $V(y^2)$, then we should see four points. This is exactly what Bezout's theorem tells us to expect, although we have to adapt our geometric intuition. L is an example of a *complete intersection*; we'll meet it again in Chapter 3.

Next, consider the ideal

$$F = \langle y^2, xy, x^2 \rangle \subseteq k[x, y, z].$$

In this case, for $i \geq 2$ a basis for $(R/F)_i$ is

$$\{z^i, z^{i-1}x, z^{i-1}y\}.$$

Bezout's theorem does not apply here, so we're on our own. First, we dehomogenize to work on the affine patch $z = 1$. A polynomial $f(x, y)$ will be in I iff f, $\partial f/\partial x$, and $\partial f/\partial y$ all vanish at $(0, 0)$. So we are imposing three conditions on the coefficients of f, and in Chapter 7 we'll see that this is the reason that $HP(R/F, i) = 3$. F is an example of a *fatpoint* ideal.

Exercise 2.3.11. Suppose a set of objects is parameterized by the points of an irreducible variety X. A condition is *generic* if it holds on a Zariski open subset of X. Show that (up to multiplication by k^*) the set of homogeneous degree two polynomials in $k[x, y, z]$ can be identified with $\mathbb{P}^5$, so a point of $\mathbb{P}^5$ corresponds to a conic curve in $\mathbb{P}^2$. Prove that a generic plane conic is *smooth* (Exercise A.3.2). $\diamond$

Supplemental reading: Hilbert functions and polynomials, Chapter 9 of Cox–Little–O'Shea [23] and Chapter 13 of Harris [52] are good references.

Chapter 3

Free Resolutions and Regular Sequences

Suppose we are handed a module M and asked to describe it. Heeding Thoreau's dictum to "Simplify, simplify!", we might begin by trying to determine if M is a direct sum of two other modules, $M \simeq M_1 \oplus M_2$. Of course, usually this will not be the case, so we should look for alternatives. A direct sum $M \simeq M_1 \oplus M_2$ gives rise to a short exact sequence

$$0 \longrightarrow M_1 \longrightarrow M \longrightarrow M_2 \longrightarrow 0.$$

Not every short exact sequence arises from a direct sum, so a reasonable substitute for a direct sum decomposition of M is an exact sequence

$$0 \longrightarrow N \longrightarrow M \longrightarrow M/N \longrightarrow 0.$$

This sequence is familiar to us from group theory, with the role of N played by a normal subgroup. We can glean lots of information from such a short exact sequence; for example, if the modules are graded (as in the last chapter), then knowing the Hilbert functions of any two modules in the sequence will tell us the Hilbert function of the third. In a nutshell, the idea is to understand an arbitrary module M by fitting it into an exact sequence with modules which we understand; for example, by fitting M into a sequence of free modules. Given such a sequence of free modules, we can compute all of the invariants of M introduced in the last chapter. Another way to obtain an exact sequence is to map a module to itself via multiplication by a fixed ring element f:

$$m \longrightarrow f \cdot m.$$

When M is $k[x_0, \ldots, x_n]/I$ and f is not a zero-divisor on M, this corresponds to slicing $V(I)$ with a generic hypersurface. This yields a geometric interpretation of the Hilbert polynomial, and leads us to study the notion of a regular sequence.

Key concepts: Free module, free resolution, Hilbert syzygy theorem, regular sequence, mapping cone.

3.1 Free Modules and Projective Modules

In a sense, free modules are the nicest of all possible modules. For example, every module is the homomorphic image of a free module – if we don't mind extravagance, we can take a generator for every element of the module. For a finitely-generated, graded, free module over a polynomial ring, we have a nice formula for the dimension of each graded piece. As it turns out, free modules fit into a somewhat broader class of modules – projective modules.

Definition 3.1.1. *An R-module P is a projective module if for any surjection of R-modules $A \xrightarrow{f} B$ and homomorphism $P \xrightarrow{g} B$ there exists a homomorphism $h : P \to A$ making the diagram below commute:*

(A diagram is called commutative if following arrows different ways gives the same result; in the above case this says $g = fh$).

The following lemma gives three different characterizations of projective modules:

Lemma 3.1.2. *The following are equivalent:*

1. *P is projective.*
2. *Every exact sequence $0 \longrightarrow N \longrightarrow M \xrightarrow{f} P \longrightarrow 0$ splits: there exists $h : P \to M$ such that fh is the identity on P.*
3. *There exists K such that $P \oplus K \simeq F$ for some free module F.*

Proof. $1 \Rightarrow 2$ is easy, and $2 \Rightarrow 3$ is basically Exercise 8.1.3. For $3 \Rightarrow 1$, take a free module F such that $P \oplus K \simeq F$ and make a commutative diagram

Since F is free, we can map F to A in a way which makes the diagram commute, but then since K goes to zero in B, it must be in the kernel of the

map from A to B. In other words, we get a map from P to A which makes the diagram commute. □

Since it is often necessary to find objects or maps which make a diagram commute, projective modules are very important. In the next exercise, you'll prove that over a local ring (a ring with a unique maximal ideal) a projective module must be free. Polynomial rings behave like local rings, for the following reason: any homogeneous ideal in a polynomial ring is necessarily contained in the ideal m generated by the variables, so m will play the role of maximal ideal. In particular ([28], Exercise 4.11), a finitely-generated, graded projective module over $k[x_1, \ldots, x_n]$ is actually a graded *free* module!

Exercise 3.1.3. The Jacobson radical $\mathcal{J}(R)$ of a ring R is the intersection of all the maximal ideals of R. For $I \subseteq \mathcal{J}(R)$ and M a finitely generated R-module, *Nakayama's lemma* tells us that $IM = M \Rightarrow M = 0$. For a local ring $\mathcal{J}(R)$ is just the maximal ideal. Use Nakayama's lemma to prove that a finitely generated projective module over a local ring is in fact free. You can check your proof in the hints and solutions section of [28] (Exercise 4.11). ◇

3.2 Free Resolutions

For the remainder of this chapter, R will denote a polynomial ring over a field. In Chapter 2 we studied the graded module R/I where $R = k[x, y, z]$, $I = \langle x^3 + y^3 + z^3 \rangle$. I being principal, it followed easily that

$$HF(R/I, i) = dim_k R_i - dim_k R_{i-3}.$$

Put another way, there is a graded exact sequence

$$0 \longrightarrow R(-3) \overset{\cdot (x^3+y^3+z^3)}{\longrightarrow} R \longrightarrow R/\langle x^3 + y^3 + z^3 \rangle \longrightarrow 0.$$

What about when we added $\langle x \rangle$ to the ideal? In that case we just used x to kill a variable and computed for a ring with two variables. But we could also write down another graded exact sequence:

$$0 \longrightarrow R(-4) \longrightarrow R(-1) \oplus R(-3) \overset{[x, x^3+y^3+z^3]}{\longrightarrow} R$$
$$\longrightarrow R/\langle x, x^3 + y^3 + z^3 \rangle \longrightarrow 0.$$

Recall that we have to shift the gradings in the above sequence in order to make the maps in the complex have degree zero. The map

$$[x, x^3 + y^3 + z^3]$$

from R^2 to R sends one generator of R^2 (call it ϵ_1) to x, and the other generator of R^2 (call it ϵ_2) to $x^3 + y^3 + z^3$. Since x is a degree one element of R, ϵ_1 must have degree one, and since $x^3 + y^3 + z^3$ is of degree three, ϵ_2 must have degree three. It is easy to check (do so!) that the kernel of the map is generated by $(x^3 + y^3 + z^3)\epsilon_1 - x\epsilon_2$. So the kernel is a free module, generated in degree four. The payoff for our careful bookkeeping is that the complex of graded modules is also exact at the level of vector spaces. We can now compute the dimension of the degree i piece of R/I as an alternating sum of the degree i pieces of the exact sequence. By Exercises 2.2.1 and 2.3.1, we know

$$HP(R/I, i) = HP(R, i) - HP(R(-1), i) - HP(R(-3), i)$$
$$+ HP(R(-4), i)$$
$$= HP(R, i) - HP(R, i-1) - HP(R, i-3) + HP(R, i-4)$$
$$= \binom{i+2}{2} - \binom{i+1}{2} - \binom{i-1}{2} + \binom{i-2}{2}$$

Exercise 3.2.1. Recall that if $I = \langle f, g \rangle \subseteq k[x, y, z]$ where f and g are homogeneous polynomials of degrees d, e *having no common factors* and k is algebraically closed, then Bezout's theorem says $V(I)$ consists of $d \cdot e$ points in $\mathbb{P}^2$. Prove this by finding a *graded* free resolution for R/I and computing $HP(R/I, i)$. $\Diamond$

The amazing fact is that we can always "approximate" a finitely generated graded module (over the polynomial ring) with a finite exact sequence of free modules (a *finite free resolution*):

Theorem 3.2.2 (Hilbert Syzygy Theorem). *If M is a finitely generated graded module over the polynomial ring $R = k[x_1, \ldots, x_n]$, then there exists a graded exact sequence of modules:*

$$0 \to F_n \longrightarrow F_{n-1} \longrightarrow \cdots \longrightarrow F_1 \longrightarrow F_0 \longrightarrow M \longrightarrow 0,$$

where the F_i are finitely generated and free.

Given a module M, the res command computes a free resolution and displays the modules (without shifts) in the resolution (M itself is not displayed). Once the resolution is in hand, you can see all the differentials by adding the suffix .dd to the name of a resolution.

```
i2 : Mr = res coker matrix {{x,x^3+y^3+z^3}}

          1      2      1
o2 = R   <-- R   <-- R

          0      1      2


o2 : ChainComplex

i3 : Mr.dd
                          1
o3 = -1 : 0 <----- R   : 0
                   0


          1                                    2
       0 : R  <-------------------- R   : 1
               {0} | x y3+z3 |


          2                                    1
       1 : R  <-------------------- R   : 2
               {1} | -y3-z3 |
               {3} | x      |
```

Notice that Macaulay 2 used x to prune the x^3 term from $x^3 + y^3 + z^3$. Printing out all the differentials can be cumbersome, because in many interesting examples the matrices can be quite large. But there is a happy medium: the command betti gives a concise encoding of the numerical information in the free resolution. If Mr is the name of a resolution, then betti Mr prints a diagram where the numbers on the top row are the ranks of the modules in the resolution. In the ith column, the number a_{ij} in the row labeled with a j : indicates that F_i has a summand of the form $R(-i-j)^{a_{ij}}$. Notice that the module itself is not displayed in the free resolution. For example, if we included shifts, the free resolution above could be written as $R \longleftarrow R(-1) \oplus R(-3) \longleftarrow R(-4)$.

```
i4: betti Mr

o4 = total: 1 2 1
        0: 1 1 .
        1: . . .
        2: . 1 1
```

Let's see the betti diagram for the twisted cubic:

```
i2 : N=res coker matrix  {{z^2- y*w, y*z - x*w,
     y^2 - x*z}}


         1       3       2
o2 = R   <-- R   <-- R   <-- 0


     0       1       2       3


o2 : ChainComplex

i3 : N.dd
             1                                   3
o3 = 0 : R   <--------------------------- R   : 1
                 | y2-xz yz-xw z2-yw |


             3                       2
     1 : R   <------------------ R   : 2
                 {2} | -z  w  |
                 {2} | y   -z |
                 {2} | -x  y  |
             2
     2 : R   <------ 0 : 3
                 0


o3: ChainComplexMap

i4 : betti N

o4 = total: 1 3 2
        0: 1 . .
        1: . 3 2
```

It is easy to see that free resolutions exist: if M is a finitely generated, graded module over R (paradigm: $M = R/I$) with generators $m_1, \ldots, m_{k_0}$, then we may define a map:

$$R^{k_0} \xrightarrow{\phi_0} M \longrightarrow 0,$$

where $\phi_0 : \epsilon_i \to m_i$. Of course, we can complete the diagram to a short exact sequence by adding the kernel K of ϕ on the left.

Lemma 3.2.3. *If R is Noetherian, then K is finitely generated.*

Proof. First, a submodule N of a Noetherian module M is Noetherian (obvious since any infinite chain of submodules of N is also an infinite chain of submodules of M). So the result holds if R Noetherian $\Rightarrow R^k$ Noetherian. This follows by induction, and the following lemma. $\square$

Lemma 3.2.4. *For a short exact sequence $0 \to M_1 \to M_2 \to M_3 \to 0$, if M_1 and M_3 are Noetherian, then so is M_2.*

Proof. Suppose we have an infinite chain of submodules of M_2. If we map them forward to M_3, we get a chain of submodules of M_3, which must stabilize, say at m_3, and if we intersect the chain with M_1, we get a chain of submodules of M_1, which must stabilize, say at m_1. Put $m_2 = max(m_1, m_3)$ and we're done. $\square$

Thus, if M is finitely generated we have an exact sequence $0 \to K \to R^{k_0} \to M \to 0$ where K is also finitely generated. So K is also the image of a free module R^{k_1}, and we can splice to obtain an exact sequence $R^{k_1} \xrightarrow{\phi_1} R^{k_0} \to M \to 0$. Repeating the process yields a free resolution of M; if at some point ϕ_i is injective then the process stops and the resolution is called *finite*. The key point of the Hilbert syzygy theorem is that a graded module M over a polynomial ring has a finite free resolution. This is simply not true over a non-polynomial ring. Over the ring $T = k[x]/\langle x^2 \rangle$, a free resolution of $\langle x \rangle$ is

$$\cdots \longrightarrow T(-2) \xrightarrow{\cdot x} T(-1) \xrightarrow{\cdot x} \langle x \rangle \longrightarrow 0.$$

A free resolution of a graded module M is called *minimal* if there are no constant terms in any of the maps; if a constant entry occurs then the maps can be pruned. For $R = k[x, y]$ the maps defined by matrices

$$\phi_1 = [x^2, yx^2] \text{ and } \phi_2 = [x^2]$$

have the same image I, but the matrices correspond to different free resolutions:

$$0 \longrightarrow R(-3) \xrightarrow{\begin{bmatrix} y \\ -1 \end{bmatrix}} R(-2) \oplus R(-3) \xrightarrow{[x^2, yx^2]} I \longrightarrow 0.$$

$$0 \longrightarrow R(-2) \xrightarrow{[x^2]} I \longrightarrow 0.$$

The point is that ϕ_1 corresponds to a non-minimal choice of generators for I. As another example, the sequence

$$0 \longrightarrow R \xrightarrow{1} R \longrightarrow 0$$

is exact, so if we graft this onto an exact sequence $\cdots \longrightarrow F_i \xrightarrow{d_i} F_{i-1} \longrightarrow \cdots$ then we obtain another exact sequence:

$$\cdots \longrightarrow F_{i+1} \longrightarrow R \oplus F_i \xrightarrow{\begin{bmatrix} 1 & 0 \\ 0 & d_i \end{bmatrix}} R \oplus F_{i-1} \longrightarrow F_{i-2} \longrightarrow \cdots.$$

The matrices in the minimal free resolution obviously cannot be unique. For example, if $I = \langle x^2, y^2 \rangle$, then in the exact sequence

$$\longrightarrow R^2(-2) \xrightarrow{\phi} I \longrightarrow 0$$

ϕ could be defined by either $[x^2, y^2]$ or $[x^2 - y^2, x^2 + y^2]$ (well, if char$(k) \neq 2$). However, the free modules which appear in a minimal free resolution *are* unique ([28], Theorem 20.2). You should prove this for yourself after working Exercise 8.2.2! We will prove the Hilbert Syzygy Theorem in Chapter 8 using the tools of homological algebra; in fact, the proof is a beautiful illustration of what the machinery of derived functors can do. For the moment, we take the theorem on faith and derive some corollaries. So, let $R = k[x_0, \ldots, x_n]$, and suppose M has free resolution

$$0 \longrightarrow F_{n+1} \longrightarrow \cdots \longrightarrow F_0 \longrightarrow M \longrightarrow 0,$$

where F_k has rank r_k and

$$F_k \simeq \bigoplus_{l=1}^{r_k} R(-a_{kl}).$$

We obtain another explanation of why the Hilbert function becomes a polynomial in large degrees – it is simply a sum of binomial coefficients:

$$HP(M, i) = \sum_{j=0}^{n+1} (-1)^j HP(F_j, i) = \sum_{j=0}^{n+1} (-1)^j \sum_{l=1}^{r_j} \binom{n + i - a_{jl}}{n}.$$

The Hilbert series for R is $1/(1 - t)^{n+1}$, so

$$HS(R(-a), t) = \frac{t^a}{(1 - t)^{n+1}}.$$

Thus

$$HS(F_k, t) = \frac{\sum_{l=1}^{r_k} t^{a_{kl}}}{(1-t)^{n+1}}.$$

This proves the assertion following Definition 2.2.4 that $HS(M, t) = P(M, t)/(1-t)^{n+1}$ where $P(M, t) \in \mathbb{Z}[t, t^{-1}]$.

3.3 Regular Sequences, Mapping Cone

In Chapter 2, we claimed that slicing with a generic hyperplane dropped the dimension of $V(I)$ by one (as long as $V(I)$ is nonempty). Of course, slicing with a generic hypersurface would yield the same result.

Exercise 3.3.1. Prove that for a homogeneous ideal $I \subseteq R$ and homogeneous polynomial f of degree d that there is a graded exact sequence

$$0 \longrightarrow R(-d)/\langle I : f \rangle \longrightarrow R/I \longrightarrow R/\langle I, f \rangle \longrightarrow 0.$$

Hint:

$$0 \longrightarrow \langle I, f \rangle / I \longrightarrow R/I \longrightarrow R/\langle I, f \rangle \longrightarrow 0$$

is clearly exact. How can you get a graded map from R to $\langle I, f \rangle / I$? What is the kernel? $\diamond$

Recall that $f \in R$ is a nonzerodivisor on M if $f \cdot m \neq 0$ for all nonzero $m \in M$; in particular, f is a nonzerodivisor on R/I iff $I : f = I$. Suppose we write

$$HP(R/I, i) = \frac{a_m}{m!} i^m + \dots$$

If f is a homogeneous linear form which is a nonzerodivisor on R/I, then from Exercise 3.3.1 we obtain

$$HP(R/\langle I, f \rangle, i) = HP(R/I, i) - HP(R/I, i-1) = \frac{a_m}{(m-1)!} i^{m-1} + \dots$$

So just as we claimed, slicing with such a hyperplane causes the dimension to drop, while preserving the degree. In fact, requiring that the hyperplane correspond to a nonzerodivisor turns out to be too restrictive. You may be thinking that this is because it is possible to have $I : f \neq I$, but nevertheless

$$\dim V(I, f) = \dim V(I) - 1.$$

For example, if $V(I)$ consists of a curve C and some isolated points, then we can choose a hyperplane $V(f)$ which does not contain any component of C

(*C* could have several pieces), but which picks up some isolated points. Then $V(I, f)$ will indeed have dimension one less than $V(I)$. The problem is that the degree will change to reflect the number of isolated points which lie on $V(f)$ – here's the simplest possible case; $V(I)$ is the union of the line $\{x = 0\}$ and the point $(1{:}0{:}0)$.

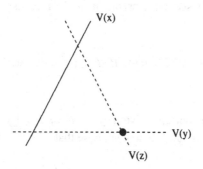

```
i1 : R=ZZ/101[x,y,z];

i2 : I=ideal(x*y,x*z);

o2 : Ideal of R

i3 : codim (I + ideal(y))

o3 = 2

i4 : degree (I + ideal(y))

o4 = 2

i5 : codim I

o5 = 1

i6 : degree I

o6 = 1
```

Of course, this computation simply reflects the fact that our chosen hyperplane $V(y)$ met $V(x)$ in one point (as required by Bezout), *but* also contained the point $(1 : 0 : 0)$. The point is that the chosen hyperplane was not generic.

The real reason that requiring a hyperplane correspond to a nonzerodivisor is too restrictive a condition is that if the primary decomposition of I has a component Q which is primary to the maximal ideal, then there can be no nonzerodivisors on R/I. But this component is geometrically irrelevant (in projective space, $V(Q)$ is empty), so we want to ignore Q. This is easily justified by an exact sequence argument as in Exercise 2.3.5. We need a lemma:

Lemma 3.3.2 (Prime Avoidance). *If* $I \subseteq \bigcup_{i=1}^{n} P_i$, *with* P_i *prime, then* $I \subseteq P_i$ *for some* i.

Proof. We prove the contrapositive: $I \nsubseteq P_i \ \forall i \Rightarrow I \nsubseteq \bigcup_{i=1}^{n} P_i$. Induct on n, the base case being trivial. We now suppose that

$$I \nsubseteq P_i \ \forall i, \text{ and } I \subseteq \bigcup_{i=1}^{n} P_i,$$

and arrive at a contradiction. From our inductive hypothesis, for each i, $I \nsubseteq \bigcup_{j \neq i} P_j$. In particular, for each i there is an x_i which is in I but is not in $\bigcup_{j \neq i} P_j$. Notice that if $x_i \notin P_i$ then $x_i \notin \bigcup_{j=1}^{n} P_j$, and we have an immediate contradiction. So suppose for every i that $x_i \in P_i$. Consider the element

$$x = \sum_{i=1}^{n} x_1 \cdots \hat{x}_i \cdots x_n.$$

By construction, $x \in I$. We claim that $x \notin \bigcup_{i=1}^{n} P_i$. To see this, observe that $x_1 \cdots \hat{x}_i \cdots x_n \notin P_i$, because for each index $k \neq i$, x_k is not in $\bigcup_{j \neq k} P_j$, so in particular is not in P_i. Since P_i is prime, this proves that $x_1 \cdots \hat{x}_i \cdots x_n \notin P_i$. But every other monomial of x is in P_i, since every other monomial contains x_i. This shows that $x \notin P_i$ for any i, hence $x \notin \bigcup_{j=1}^{n} P_j$, a contradiction. $\square$

With prime avoidance in hand, we can finish our justification that the Hilbert polynomial does indeed give the "geometrically correct" values for dimension and degree. If I has a primary component primary to the maximal ideal, we can remove it, since it is irrelevant for computing the Hilbert polynomial. Call the new ideal I'. By prime avoidance, the union of the associated primes of I' cannot contain the maximal ideal, so there must be a (homogeneous) linear form f not in the union of the associated primes. By Lemma 1.3.8, f is a nonzerodivisor on R/I', and we have found a way to chop down the dimension of $V(I')$, and hence $V(I)$, while preserving the degree. Notice that when we add f to I, all the primary components will increase codimension. We can repeat the process exactly $\dim(V(I))$ times, and at the end we are left with an ideal with Hilbert polynomial a constant equal to the degree of $V(I)$.

We return now to the situation $I : f = I$, which we saw was synonymous with saying that f is a nonzerodivisor on R/I and hence $V(I, f)$ has dimension one less than $V(I)$. It is natural to start from scratch with a single polynomial, and iterate this process; while we're at it, we generalize from R/I to an arbitrary module. For a graded module M, we define a *regular sequence* on M as a sequence of (non-constant) homogeneous polynomials

$$\{f_1, f_2, \ldots, f_m\},$$

such that f_1 is a nonzerodivisor on M, and f_i is a nonzerodivisor on $M/\langle f_1, \ldots, f_{i-1}\rangle M$. For example if $R = k[x, y]$, then $\{x, y\}$ is a regular sequence on R, since x is a nonzerodivisor on R and y is a nonzerodivisor on $R/\langle x \rangle \simeq k[y]$. It is easy to see that in a polynomial ring R, the variables (actually, any subset of the variables) are a regular sequence on R. See if you can find some other regular sequences on R (hint: what about powers of variables). Check what happens when you compute a free resolution for an ideal generated by a regular sequence. What you should see is that the only relations that occur in the free resolution are the obvious ones, for example, if $I = \{f_1, f_2, f_3\}$ is a regular sequence on R, the free resolution of R/I will be (in Macaulay 2 format):

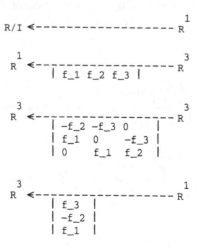

An ideal generated by a regular sequence on R is called a *complete intersection*. One way of stating Bezout's theorem is to say that the ideal generated by two polynomials with no common factor is a complete intersection. Let's see what happens for a regular sequence of three polynomials in $\mathbb{P}^3$. Suppose the complete intersection shown above is sitting in $R = k[x_0, \ldots, x_3]$, with degree $f_i = d_i$. Thus, the Hilbert polynomial of R/I is

$$dim_k R_i - \sum_{j=1}^{3} dim_k R(-d_j)_i + \sum_{1 \le j < k \le 3} dim_k R(-d_j - d_k)_i$$
$$-dim_k R(-d_1 - d_2 - d_3)_i.$$

Grinding through the computation, we obtain

$$HP(R/I, i) = d_1 \cdot d_2 \cdot d_3.$$

This means that the hypersurfaces defined by the three polynomials intersect in $d_1 \cdot d_2 \cdot d_3$ points, so each polynomial chops dimension down by one. This is where the terminology complete intersection comes from: we go from the surface in $\mathbb{P}^3$ defined by $V(f_1)$ to the curve defined by $V(f_1, f_2)$ to the set of points $V(f_1, f_2, f_3)$. We can use Exercise 3.3.1 to obtain a free resolution for an ideal generated by a regular sequence $\{f_1, f_2, f_3\}$ as follows: first, let $I = \langle f_1, f_2 \rangle$. By Exercise 3.2.1 we know what a free resolution for R/I looks like:

$$0 \longrightarrow R(-d_1 - d_2) \longrightarrow R(-d_1) \oplus R(-d_2) \longrightarrow R \longrightarrow R/I \longrightarrow 0.$$

Since f_3 is a nonzerodivisor on R/I, this means that $\langle I : f_3 \rangle$ is simply I, so that

$$R(-d_3)/\langle I : f_3 \rangle = R(-d_3)/I.$$

Thus, just shifting the degree of the free resolution of R/I gives us a free resolution for $R(-d_3)/\langle I : f_3 \rangle$. We have a diagram:

$$
\begin{array}{ccccccccc}
 & & & & & & & 0 & \\
 & & & & & & & \downarrow & \\
0 \to & R\left(-\sum_{i=1}^{3} d_i\right) & \xrightarrow{\begin{bmatrix} f_2 \\ -f_1 \end{bmatrix}} & \begin{array}{c} R(-d_1 - d_3) \\ \oplus \\ R(-d_2 - d_3) \end{array} & \xrightarrow{[f_1, f_2]} & R(-d_3) & \to & R(-d_3)/I & \to 0 \\
 & & & & & & & \downarrow \cdot f_3 & \\
0 \to & R(-d_1 - d_2) & \xrightarrow{\begin{bmatrix} f_2 \\ -f_1 \end{bmatrix}} & \begin{array}{c} R(-d_1) \\ \oplus \\ R(-d_2) \end{array} & \xrightarrow{[f_1, f_2]} & R & \to & R/I & \to 0 \\
 & & & & & & & \downarrow \pi & \\
 & & & & & & & R/\langle I, f_3 \rangle & \to 0 \\
 & & & & & & & \downarrow & \\
 & & & & & & & 0 &
\end{array}
$$

It is obvious how to make a vertical map ϕ_0 from $R(-d_3)$ to R so that the rightmost square above commutes: ϕ_0 is just multiplication by f_3. In fact, by defining vertical maps further back in the resolution as f_3 times the identity map, we can make each square commute. Let's call these vertical maps ϕ_i. For simplicity, rename the horizontal maps ψ_i and ξ_i. We have:

$$
\begin{array}{ccccccccc}
& & & & & & & 0 & \\
& & & & & & & \downarrow & \\
F: 0 \to R\left(-\sum_{i=1}^{3} d_i\right) & \xrightarrow{\psi_2} & \begin{array}{c} R(-d_1 - d_3) \\ \oplus \\ R(-d_2 - d_3) \end{array} & \xrightarrow{\psi_1} & R(-d_3) & \xrightarrow{\psi_0} & R(-d_3)/I & \to 0 \\
\phi_2 \downarrow & & \phi_1 \downarrow & & \phi_0 \downarrow & & \cdot f_3 \downarrow & \\
& & R(-d_1) & & & & & \\
G: 0 \to R(-d_1 - d_2) & \xrightarrow{\xi_2} & \oplus & \xrightarrow{\xi_1} & R & \xrightarrow{\xi_0} & R/I & \to 0 \\
& & R(-d_2) & & & & & \\
& & & & & & \pi \downarrow & \\
& & & & & & R/\langle I, f_3 \rangle & \to 0 \\
& & & & & & \downarrow & \\
& & & & & & 0 & \\
\end{array}
$$

How do we get a free resolution for $R/\langle I, f_3 \rangle$? You might think that we could just take the cokernels of the vertical maps, but a moment of thought shows that it is not this simple. Notice we have a map from $G_0 \simeq R$ onto $R/\langle I, f_3 \rangle$ via $\pi \circ \xi_0$; the kernel of this map is generated by the images of ϕ_0 and ξ_1. Thus, there is a map from $F_0 \oplus G_1$ to $G_0 = R$ which is the start of a free resolution for $R/\langle I, f_3 \rangle$.

Exercise 3.3.3. Verify that the kernel of $\pi \circ \xi_0$ is as claimed. Show that there is a complex

$$
F_{i-1} \oplus G_i \xrightarrow{\partial_i} F_{i-2} \oplus G_{i-1},
$$

with $\partial_0 = \pi \circ \xi_0$ mapping G_0 to $R/\langle I, f_3 \rangle$, ∂_1 mapping $F_0 \oplus G_1$ to G_0 via

$$
\partial_1 = [\phi_0 \quad \xi_1]
$$

and for $i > 1$:

$$
\partial_i = \begin{bmatrix} \psi_{i-1} & 0 \\ (-1)^{i-1}\phi_{i-1} & \xi_i \end{bmatrix}
$$

This construction is an instance of a *mapping cone*. If the f_i are a regular sequence prove that the resulting complex is actually exact. In particular, this gives a resolution for the residue field of a polynomial ring. Even if the f_i are not a regular sequence we still obtain a complex, called the *Koszul complex*.

If this was too easy, try writing down a mapping cone resolution for the short exact sequence of Exercise 3.3.1 when $I : f \neq I$. For more details or a hint if you get stuck, see Eisenbud A3.12. ◇

So, now we know how to find the free resolution of a complete inter-section. We'll prove shortly that the length of a maximal regular sequence in $k[x_1, \ldots, x_n]$ is n, and that n generic homogeneous polynomials form a regular sequence.

Example 3.3.4. What happens if we take more than n such polynomials, i.e. what is the free resolution for $I = \langle f_1, \ldots, f_k \rangle \subseteq k[x_1, \ldots, x_n]$ when $k > n$ and the f_i are generic and homogeneous? If $n > 3$, this is an open research problem! Go to

http://www.ams.org/mathscinet,

(you'll probably need to be on a university machine) and do a search on Ralf Fröberg to learn more – Fröberg made a conjecture about this prob-lem in [39] and proved the conjecture in the case $n = 2$. Let's code it up in Macaulay 2! Open a file called, e.g., "randform" and type in the little script below:

```
Resrandform=(n,l)->(
    R=ZZ/31991[x_1..x_n];
    I=ideal(0_R);
    scan(l,i->(I=I+ideal random(R^{i},R^1)));
    print betti res coker mingens I)
--script to print resolution of random forms.
--Input is n=number of variables and l=a list
--of the degrees of the forms.
```

The script starts by creating a ring in the specified number of variables. Then it loops through the list l of degrees, generating a random form of the specified degree for each list element, and adding it to the ideal. When the loop ends, it prints out the betti diagram of the ideal.

Of course, there is never just one way to code a problem. For example, we could cut out the scan loop with the following syntax:

```
Resrandform2=(n,l)->(R=ZZ/31991[x_1..x_n];
                I=ideal random(R^l,R^1);
                print betti res coker mingens I)
```

We now try the code:

```
i1: load "randform";

i2 : Resrandform(4,{6,6,6,6,7,7,7,7})
total: 1 8 105 164 66
    0: 1 .   .   . .
    1: . .   .   . .
    2: . .   .   . .
    3: . .   .   . .
    4: . .   .   . .
    5: . 4   .   . .
    6: . 4   .   . .
    7: . .   .   . .
    8: . .   .   . .
    9: . .   .   . .
   10: . . 105 164 66
```

Exercise 3.3.5. When you type `Resrandform(3,{2,2,2,2,2,2})` and `Resrandform(3,{2,2,2,2,2,2,2})`, you get the same output. Why is this to be expected? ◇

Supplemental reading: For free resolutions, see Chapter 19 of Eisenbud [28] or Chapter 5 of the second Cox–Little–O'Shea book [24]. For more on the current status of the problem on random forms, go to

```
http://xxx.lanl.gov/archive/math
```

or

```
http://front.math.ucdavis.edu/
```

and check out the paper of Migliore and Miro-Roig [66]; another recent preprint is that of Pardue and Richert [76], available at the authors' website.

Chapter 4

Gröbner Bases and the Buchberger Algorithm

This chapter gives a "look under the hood" at the algorithm that actually lets us perform computations over a polynomial ring. In order to work with polynomials, we need to be able to answer the ideal membership question. For example, there is no chance of writing down a minimal free resolution if we cannot even find a minimal set of generators for an ideal. How might we do this? If $R = k[x]$, then the Euclidean algorithm allows us to solve the problem. What makes things work is that there is an invariant (degree), and a process which reduces the invariant. Then ideal membership can be decided by the division algorithm. When we run the univariate division algorithm, we "divide into" the initial (or lead) term. In the multivariate case we'll have to come up with some notion of initial term – for example, what is the initial term of $x^2y + y^2x$? It turns out that this means we have to produce an ordering of the monomials of $R = k[x_1, \ldots, x_n]$. This is pretty straightforward. Unfortunately, we will find that even once we have a division algorithm in place, we still cannot solve the question of ideal membership. The missing piece is a multivariate analog of the Euclidean algorithm, which gave us a good set of generators (one!) in the univariate case. But there is a simple and beautiful solution to our difficulty; the Buchberger algorithm is a systematic way of producing a set of generators (a *Gröbner basis*) for an ideal or module over R so that the division algorithm works. The Buchberger algorithm and Gröbner bases are covered with wonderful clarity in the book [23] of Cox, Little, and O'Shea, so the treatment here is terse. We study three key byproducts of the Buchberger algorithm: computation of numerical invariants by passing to the initial ideal, Gröbner bases for modules and computation of syzygies, and determination of the equations of the projection of a variety (elimination).

Key concepts: Monomial order, initial ideal, Gröbner basis, Buchberger algorithm, elimination, computation of syzygies.

4.1 Gröbner Bases

In order to decide ideal membership, we want a division algorithm, so we need to order the monomials of R. An order $>$ on a set S is a *total order* if for two elements $\alpha, \beta \in S$, one and only one of the following possibilities occurs: $\{\alpha > \beta, \alpha < \beta, \alpha = \beta\}$.

Definition 4.1.1. *Associate to a monomial* $x_1^{\alpha_1} \cdots x_n^{\alpha_n}$ *the exponent vector* $\alpha = (\alpha_1, \ldots, \alpha_n)$. *A monomial order on* $R = k[x_1, \ldots, x_n]$ *is a total order on n-tuples of nonnegative integers, which also satisfies*

1. *For any* γ, *if* $\alpha > \beta$, *then* $\alpha + \gamma > \beta + \gamma$.
2. *Any nonempty subset has a smallest element* ($>$ *is a well-ordering*).

Example 4.1.2. (Examples of monomial orders): Let $|\alpha| = \sum \alpha_i$

1. Pure Lexicographic (the phone book order): $\alpha > \beta$ if the leftmost nonzero entry of $\alpha - \beta$ is positive.
2. Graded Lexicographic: $\alpha > \beta$ if $|\alpha| > |\beta|$ or $|\alpha| = |\beta|$ and the leftmost nonzero entry of $\alpha - \beta$ is positive.
3. Graded Reverse Lexicographic: $\alpha > \beta$ if $|\alpha| > |\beta|$ or $|\alpha| = |\beta|$ and the rightmost nonzero entry of $\alpha - \beta$ is negative.

For example, in $k[x, y, z]$, we associate the vector $(1, 0, 0)$ to the monomial x and the vector $(0, 1, 2)$ to the monomial yz^2. Then in pure lex, $x > yz^2$ since $(1, 0, 0) - (0, 1, 2) = (1, -1, -2)$ has leftmost nonzero entry positive. In graded lex, $yz^2 > x$ since $|x| = 1$ but $|yz^2| = 3$.

Exercise 4.1.3. Order the variables x, y, z with respect to the above orders. Do the same for the monomials

$$x^2, y^2, z^2, xy, xz, yz. \quad \diamond$$

Assume we have chosen a monomial order on R, and write $f \in R$ as

$$f = \sum_{c_\alpha \neq 0} c_\alpha x^\alpha.$$

The initial monomial in(f) of f is the largest (with respect to our chosen order) x^α which appears in this expression. If we are not working over a field, then it is important to distinguish between in(f) and the initial term of f, which is $c_\alpha x^\alpha$, with $x^\alpha = $ in(f). Since we will always work over a field, we can be cavalier about this distinction, because we can always assume f is monic. Given $f \in R$ and set of polynomials

$$\{f_1, \ldots, f_m\},$$

we can run a division algorithm to try to write f as a combination of $f_1, \ldots, f_m$.

Algorithm 4.1.4. The Division Algorithm

```
div := 0;
rem := 0;
While f <> 0 do
        if in(f_i)*a_i =  in(f) then{
            div := div + a_i*f_i;
            f :=f-a_i*f_i}
        else{
            rem := rem + in(f);
            f :=f-in(f)}
```

So, we can write

$$f = \sum a_i \cdot f_i + r,$$

where no monomial of r is divisible by any $\text{in}(f_i)$. Of course, this does not produce unique output, because at any step, we might have several different f_i whose initial terms divide the initial term of f. On the other hand, the algorithm does terminate, because at each step, the initial term of f decreases, and by the well ordering property, such a process cannot go on indefinitely. Suppose we want to decide if

$$x^2 - y^2 \in \langle x^2 + y, xy + x \rangle.$$

Using graded lex order, we divide by $x^2 + y$, and obtain a remainder of $-y^2 - y$. Since neither x^2 nor xy divides y^2, we're done. The problem is that

$$x^2 - y^2 = x(xy + x) - y(x^2 + y).$$

So we seem to have failed miserably. But we can salvage something – notice that the reason the division algorithm didn't work is that the polynomial $x^2 - y^2$ can be written as a combination of the two generators by canceling out their leading terms. So of course in a situation like this the division algorithm is inadequate. To remedy this, we may as well enlarge our generating set to include all polynomials that arise from canceling initial terms. At first it seems bad to enlarge a generating set, but this is really what makes things work!

Definition 4.1.5 (Gröbner basis). *A subset $\{g_1, \ldots, g_n\}$ of an ideal I is a Gröbner basis for I if the ideal generated by initial monomials of elements of I (denoted $in(I)$) is generated by $\{in(g_1), \ldots, in(g_n)\}$.*

It is easy to show (do so!) that if G is a Gröbner basis for I then $\langle G \rangle = I$.

Definition 4.1.6 (Syzygy pairs). *For f, g monic polynomials, put*

$$S(f, g) := \frac{LCM(in(f), in(g))}{in(f)} \cdot f - \frac{LCM(in(f), in(g))}{in(g)} \cdot g$$

Theorem 4.1.7. *G is a Gröbner basis iff $S(g_i, g_j)$ reduces to zero mod G for all pairs $g_i, g_j \in G$.*

Proof. $\Rightarrow$ is obvious. We sketch $\Leftarrow$. For a detailed proof see [23]. Let $f \in I$ and suppose we have a set $\{g_1, \ldots, g_k\}$ where all S-pairs reduce to zero. We want to show that $in(f) \in \langle in(g_1), \ldots, in(g_k) \rangle$. We know we can write

$$f = \sum a_i g_i.$$

When we do this, two things can happen. Either $in(f)$ is the initial monomial of one of the $a_i g_i$ (in which case we're done), or we had some cancellation of initial terms. But cancellation means we had an S-pair, and since S-pairs reduce to zero, we can replace the S-pair with a new combination of the g_i. $\square$

Algorithm 4.1.8. The Buchberger Algorithm
```
Input: G:={f_1,...,f_k}
Repeat
     G':= G
     for each pair p,q in G' do
          {S := reduce S(p,q) mod G';
           if S <> 0 then G:= G + S}
Until G' = G
```

Notice that if we add elements on a pass through the for loop, then we make the ideal generated by G larger: $in(G') \subsetneq in(G)$. But since the polynomial ring is Noetherian, there are no infinite ascending chains of ideals. Hence, the algorithm must terminate!

Example 4.1.9. We compute a Gröbner basis of $\langle x^2 - y^2, xy - 1 \rangle$ with respect to lex order with $x > y$ (call the generators f_1, f_2). The first S-pair computation yields

$$S(f_1, f_2) = y(x^2 - y^2) - x(xy - 1) = x - y^3 = f_3.$$

Notice that the initial term of this polynomial is x (we're using lex!), so we cannot reduce it mod f_1, f_2. On the next pass through the loop, we compute

$$S(f_1, f_3) = 1(x^2 - y^2) - x(x - y^3) = xy^3 - y^2,$$

which does reduce to zero (by running the division algorithm with f_1, f_2, f_3 as potential divisors), and

$$S(f_2, f_3) = 1(xy - 1) - y(x - y^3) = y^4 - 1 = f_4,$$

which we cannot reduce. On the final pass, we obtain S-pairs which all reduce to zero (running the division algorithm with f_1, f_2, f_3, f_4 as potential divisors), so we're done.

You may have noticed that f_1 appeared with coefficient 1 in the computation of $S(f_1, f_3)$. Combined with the fact that $S(f_1, f_3)$ reduces to zero, this means that the generator f_1 is superfluous in the Gröbner basis. A similar argument shows that f_2 is also redundant. A Gröbner basis $\{g_1, \ldots, g_k\}$ is called *minimal* if for all i, $\langle \text{in}(g_1), \ldots, \text{in}(g_{i-1}), \text{in}(g_{i+1}), \ldots, \text{in}(g_k) \rangle \neq \langle \text{in}(g_1), \ldots, \text{in}(g_k) \rangle$.

Exercise 4.1.10. The fastest order (generally, see [12] for details) is graded reverse lex, which is the default order in Macaulay 2. Find (by hand) a Gröbner basis for the ideals $\langle x^2 + y, xy + x \rangle$ and $\langle x + y + z, xy + xz + yz, xyz \rangle$, using the three orders we've defined. Then check your work in Macaulay 2. $\diamond$

```
--set order to graded lex
i1 : R=QQ[x,y, MonomialOrder=>GLex];

i2 : gb ideal(x^2+y,x*y+x)

o2 = | y2+y xy+x x2+y |
```

Exercise 4.1.11. [24] Let X be a finite subset of d points in $\mathbb{Q}^n$. Show that after a general linear change of coordinates, we may assume that the nth

coordinates of these points are distinct. In this situation show that the Lexicographic Gröbner basis for $I(X)$ is of the form

$$\langle x_1 - p_1(x_n), x_2 - p_2(x_n), \ldots, x_{n-1} - p_{n-1}(x_n), p_n(x_n) \rangle,$$

where p_n is of degree d and the other p_i are all of degree at most $d - 1$. Hint: Lagrange interpolation. Here is an example for the points

$$\{(0, 0), (1, 1), (3, 2), (5, 3)\} \in \mathbb{Q}^2 :$$

```
i1 : R=QQ[x,y, MonomialOrder=>Lex];

i2 : I=gb intersect(ideal(x,y),ideal(x-1,y-1),
                     ideal(x-3,y-2), ideal(x-5,y-3))

o2 = | y4-6y3+11y2-6y x+1/6y3-y2-1/6y |

o2 : GroebnerBasis
```

Do we really need $\mathbb{Q}$? ◇

4.2 Monomial Ideals and Applications

In Chapter 2, we saw that the Hilbert polynomial of a finitely generated, graded R-module M could be computed by finding a free resolution of M. This is actually a very bad way to solve the problem; the right (fastest!) way to compute the Hilbert polynomial is via Gröbner bases. We now do this in the case where $M = R/I$, but first we need three little lemmas:

Lemma 4.2.1 (Macaulay). *Let R be a polynomial ring and I a homogeneous ideal. Then the Hilbert function of I is the same as the Hilbert function of $in(I)$.*

Proof. For a graded piece I_i of the ideal we have a vector space basis given by $\{f_1, \ldots, f_j\}$. We can assume that $in(f_1) > in(f_2) \ldots > in(f_j)$, so the $in(f_j)$ are linearly independent. If they don't span $in(I)_i$, pick $m \in in(I)_i$ which is not in the span, such that $in(m) = m'$ is minimal with respect to the term order. Of necessity, there is a polynomial $g \in I$ with $in(g) = m'$. Since g is a k-linear combination of the f_j, m' must be one of the $in(f_j)$ (since no cancellation can occur), contradicting the choice of m' minimal. Notice that choice of monomial order is irrelevant. □

Lemma 4.2.2. *Let $I = \langle x^{a_1}, \dots, x^{a_j} \rangle$ be a monomial ideal, and x^a a monomial. Then $x^a \in I \iff x^{a_i}$ divides x^a for some i.*

Proof. $\Leftarrow$ is obvious. For $\Rightarrow$, if $x^a \in I$, then $x^a = \sum f_j x^{a_j}$. Since each term on the right hand side is divisible by an x^{a_i}, so is each term on the left hand side. $\square$

Lemma 4.2.3. *Let $I = \langle x^{a_1}, \dots, x^{a_j} \rangle$ be a monomial ideal, and x^a a monomial. Then*

$$\left\langle \frac{x^{a_1}}{GCD(x^{a_1}, x^a)}, \dots, \frac{x^{a_j}}{GCD(x^{a_j}, x^a)} \right\rangle = I : x^a.$$

Proof. Obviously we have $\subseteq$. The other containment follows from the previous lemma. $\square$

By Macaulay's lemma, $HP(R/I, i) = HP(R/\text{in}(I), i)$. So in order to compute $HP(R/I, i)$, we compute a Gröbner basis and work with $\text{in}(I)$. If I is a monomial ideal, and $f \notin I$ is a monomial of degree d, then $I : f$ and I both have fewer generators than $\langle I, f \rangle$. This means we can compute the Hilbert polynomial inductively using Lemma 4.2.3 and the exact sequence

$$0 \longrightarrow R(-d)/(I : f) \overset{\cdot f}{\longrightarrow} R/I \longrightarrow R/\langle I, f \rangle \longrightarrow 0.$$

This method is also fast, see [11] or [14]. Let's work an example by hand:

Example 4.2.4. For $R = k[x, y, z, w]$, check that

$$I = \langle yz - xw, z^2 - yw, y^2 - xz \rangle$$

is a Gröbner basis in graded reverse lex order, so

$$\text{in}(I) = \langle y^2, yz, z^2 \rangle.$$

Since

$$\langle y^2, z^2 \rangle : yz = \langle y, z \rangle,$$

the exact sequence

$$0 \longrightarrow R(-2)/\langle y, z \rangle \overset{\cdot yz}{\longrightarrow} R/\langle y^2, z^2 \rangle \longrightarrow R/\langle y^2, yz, z^2 \rangle \longrightarrow 0.$$

shows that

$$HP(R/I, i) = HP(R/\langle y^2, z^2 \rangle, i) - HP(R(-2)/\langle y, z \rangle, i).$$

Obviously $\langle y^2 \rangle : z^2 = \langle y^2 \rangle$, so $HP(R/\langle y^2, z^2 \rangle, i) = 4i$ and $HP(R(-2)/\langle y, z \rangle, i) = i - 1$. Putting the pieces together, we have

$$HP(R/I, i) = 4i - (i - 1) = 3i + 1.$$

Let's double check that Macaulay 2 is computing the Hilbert polynomial intelligently, rather than using a free resolution. If you precede a Macaulay 2 command with time, you will get the result of the computation, as well as the time it took.

```
i1 : R=ZZ/101[x_1..x_16];
```

```
i2 : genericMatrix(R,x_1,4,4)
```

```
o2 = {0} | x_1 x_5 x_9  x_13 |
     {0} | x_2 x_6 x_10 x_14 |
     {0} | x_3 x_7 x_11 x_15 |
     {0} | x_4 x_8 x_12 x_16 |

                 4       4
o2 : Matrix R  <--- R
```

```
i3 : I = minors(2,o2);
```

```
o3 : Ideal of R
```

```
i4 : time hilbertPolynomial coker gens I
     -- used 0.05 seconds

o4 = - P  + 12*P  - 30*P  + 20*P
        3       4       5       6

o4 : ProjectiveHilbertPolynomial
```

```
i5 : time res coker gens I
     -- used 0.77 seconds

          1       36      160      315      388
o5 = R  <-- R   <-- R   <-- R   <-- R    ......

          0       1       2       3       4
```

We close this section with a fun example/exercise, which shows that for monomial ideals there are very nice descriptions of ideal theoretic properties:

Example 4.2.5. A monomial ideal I is irreducible iff it is generated by powers of the variables.

Proof. First, suppose I is irreducible, but not generated by pure powers. Let $\{m, m_1, \ldots, m_k\}$ be a set of minimal monomial generators for I, and suppose $m = x_1^{a_1} x_2^{a_2} \cdots x_n^{a_n}$ is such that no $x_i^{b_i} \in I$ divides m. Without loss of generality we may assume $a_1 \neq 0$. Put $I_1 = \langle m_1, \ldots, m_k, x_1^{a_1} \rangle$ and $I_2 = \langle m_1, \ldots, m_k, m/x_1^{a_1} \rangle$. I is properly contained in both I_1 and I_2, and is equal to the intersection, which contradicts the hypothesis that I is irreducible. Now suppose I is generated by pure powers $x_1^{a_1}, x_2^{a_2}, \ldots, x_n^{a_n}$, and that $I = I_1 \cap I_2$. Choose monomials $n_1 = x_1^{c_1} \cdots x_n^{c_n} \in I_1$ but not in I, $n_2 = x_1^{d_1} \cdots x_n^{d_n} \in I_2$ but not in I. Since $I_1 I_2 \subseteq I$, $n_1 n_2 \in I$, and since $n_1, n_2 \notin I$, $c_i, d_i < a_i$. But $x_1^{max(c_1, d_1)} \cdots x_n^{max(c_n, d_n)} \in I$, so some $x_i^{a_i}$ divides it, hence also divides either n_1 or n_2, a contradiction. $\square$

Exercise 4.2.6. For a monomial ideal I, prove

1. I is prime iff I can be generated by a subset of the variables.
2. I is radical iff I can be generated by square-free monomials.
3. I is primary iff every variable which appears in some minimal monomial generator of I also appears in I as a pure power. $\diamond$

4.3 Syzygies and Gröbner Bases for Modules

In Chapter 3, we saw that if M is a finitely generated module over a polynomial ring with M generated by elements $m_1, \ldots, m_j$, then we could define a map:

$$R^j \xrightarrow{\phi} M \longrightarrow 0,$$

which was the beginning of a free resolution for M. To actually write down the entire free resolution, we need to find the kernel of this map, and then iterate the process. Gröbner bases allow us to do this! For $M \subseteq R^p$, we may write each generator m_i as $\sum_{i=1}^p h_i \epsilon_i$. So to define the initial term of m_i, we need to order the ϵ_i, and we can (almost) treat them like another set of variables. Then to compute a Gröbner basis for M, we take pairs m_i, m_j whose initial term involves the same basis element ϵ_k, and form the syzygy pair, repeating until all syzygy pairs reduce to zero. The Buchberger algorithm and criterion may be applied exactly as in the ideal case. If you object to the condition that

$M \subseteq R^p$, just stop for a moment and notice that a finitely generated module M is determined *exactly* like a finitely generated group – by generators and relations. So a finitely generated module M with n generators is given by a sequence:

$$R^m \xrightarrow{\phi} R^n \longrightarrow M \longrightarrow 0,$$

where the columns of ϕ generate all relations on the generators of M (Note: we proved last chapter that the kernel of ϕ is Noetherian, hence finitely generated).

Algorithm 4.3.1. (Buchberger algorithm for modules) Suppose $M \subseteq R^p$, where R^p has basis $\epsilon_1, \ldots, \epsilon_p$. For pairs i, j such that $\text{in}(m_i) = f_i \epsilon_k$, $\text{in}(m_j) = f_j \epsilon_k$, put

$$m_{ij} = \frac{LCM(f_i, f_j)}{f_j}.$$

Write

$$m_{ji} m_i - m_{ij} m_j = \sum a_k m_k + r$$

(i.e. kill initial entries, and reduce what is left as much as possible by the other m_i). Then the m_i are a Gröbner basis for M iff all the r are zero.

It is worth emphasizing that there is nothing fancy going on here – this is simply the usual Buchberger algorithm, where we only form S-pairs for those generators of M whose initial entries appear in the same position. If we choose a kind of "induced order" on the ϵ_i, then the Buchberger algorithm also gives us the syzygies on M *for free*!

Theorem 4.3.2 (Schreyer). *Suppose $g_1, \ldots, g_n$ is a Gröbner basis for M. Let R^n have basis $e_1, \ldots, e_n$ and define an order on R^n via $a_i e_i > a_j e_j$ if $\text{in}(a_i g_i) > \text{in}(a_j g_j)$ in the order on M, or if $\text{in}(a_i g_i) = \text{in}(a_j g_j)$ and $i < j$. Then the elements*

$$m_{ji} e_i - m_{ij} e_j - \sum a_k e_k$$

are a Gröbner basis for the syzygies on M with respect to the above order on R^n.

The proof of this theorem is not too bad but a bit lengthy; a good reference for this and other aspects of Gröbner bases for modules is [28].

Example 4.3.3. (The twisted cubic, again) The generators of $I = \langle y^2 - xz,$ $yz - xw, z^2 - yw \rangle = \langle f_1, f_2, f_3 \rangle$ are a Gröbner basis with respect to graded reverse lex order, and the initial monomials are (respectively) $\{y^2, yz, z^2\} = \{\text{in}(f_1), \text{in}(f_2), \text{in}(f_3)\}$. I is a submodule of R^1, so we compute:

$$m_{12} = \frac{LCM(y^2, yz)}{yz} = y \quad m_{21} = \frac{LCM(yz, y^2)}{y^2} = z$$

$$m_{13} = \frac{LCM(y^2, z^2)}{z^2} = y^2 \quad m_{31} = \frac{LCM(z^2, y^2)}{y^2} = z^2$$

$$m_{23} = \frac{LCM(yz, z^2)}{z^2} = y \quad m_{32} = \frac{LCM(z^2, yz)}{yz} = z$$

Now, we form

$$m_{21}m_1 - m_{12}m_2 = z(y^2 - xz) - y(yz - xw) = xyw - xz^2 = -x(z^2 - yw),$$

which yields a syzygy

$$ze_1 - ye_2 + xe_3.$$

Exercise 4.3.4. Finish the previous example by computing the entire free resolution of I. You'll need to compute the other two potential first syzygies (one will be redundant), and then show there are no second syzygies. Now try another example – verify that the Koszul complex is a resolution of the ideal $\langle x, y, z \rangle \subseteq k[x, y, z]$. ◇

4.4 Projection and Elimination

One way to approach solving systems of equations is to project down to a lower dimensional space, find solutions there, and then try to extend them; in linear algebra this is familiar as Gaussian elimination. A very similar idea works when solving systems of polynomials, and Gröbner bases play a central role.

Definition 4.4.1. *Let* $I \subseteq k[x_1, \ldots, x_n] = R$. *The* m^{th} *elimination ideal* $_mI$ *is* $I \cap k[x_{m+1}, \ldots, x_n]$.

To understand why we need this, let's consider some geometry. Write π_m for the projection:

$$\mathbb{A}^n \xrightarrow{\pi_m} \mathbb{A}^{n-m},$$

$$\pi_m(a_1, \ldots, a_n) = (a_{m+1}, \ldots, a_n).$$

The projection of a variety need not be a variety, but it has a Zariski closure which is a variety. Here is what the definition buys us:

Theorem 4.4.2. *If k is algebraically closed, then $\overline{\pi_m(V(I))} = V(_mI)$.*

Proof. Let $(a_{m+1}, \ldots, a_n) \in \pi_m(V(I))$, $(a_{m+1}, \ldots, a_n) = \pi_m(a_1, \ldots, a_n)$ for $(a_1, \ldots, a_n) \in V(I)$. For $f \in {}_mI$, since $f \in I$, $f(a_1, \ldots, a_n) = 0$. But $f \in k[x_{m+1}, \ldots, x_n]$ so $0 = f(a_1, \ldots, a_n) = f(a_{m+1}, \ldots, a_n) = f(\pi_m(a_1, \ldots, a_n))$. Hence $\pi_m(V(I)) \subseteq V(_mI)$. For the other containment, $g \in I(\pi_m(V(I)))$, regarded as an element of R, vanishes on $V(I)$. By the Nullstellensatz $g^p \in I$. Since $g^p \in k[x_{m+1}, \ldots, x_n]$, we see that $g^p \in {}_mI$. So $g \in \sqrt{_mI}$ and $I(\pi_m(V(I))) \subseteq I(V(_mI))$. Passing to varieties yields $V(_mI) \subseteq \overline{\pi_m(V(I))}$. $\square$

Well, this does us no particular good unless we can find $_mI$. And naturally, Gröbner bases ride to the rescue:

Theorem 4.4.3. *Let $I \subseteq k[x_1, \ldots, x_n]$, and let $G = \{g_1, \ldots, g_j\}$ be a Gröbner basis for I with respect to lex order, $x_1 > x_2 > \ldots > x_n$. Then $_mG := G \cap k[x_{m+1}, \ldots, x_n]$ is a Gröbner basis for $_mI$.*

Proof. It is immediate that $_mG \subseteq {}_mI$. Take $f \in {}_mI$ and write it as

$$f = \sum h_i g_i.$$

If some g_l which appears in this expression is not in $_mI$, then $\text{in}(g_l)$ is not in $_mI$ (this is where lex is used!), so when we run the division algorithm to write f as a combination of the g_i, g_l will not appear, contradiction. Thus

$$_mI = \langle _mG \rangle.$$

To see that $_mG$ is actually a Gröbner basis, we only need to check that the S-pairs reduce to zero, but this is automatic since the original S-pairs reduced to zero using the division algorithm *with lex order*. $\square$

Exercise 4.4.4. Describe what is happening in Exercise 4.1.11 in terms of projection. Can you see why distinct n^{th} coordinates were necessary? ◇

Example 4.4.5. A very important application of elimination consists of finding the equations of the image of a polynomial map. Consider the affine version of the twisted cubic. Map

$$\mathbb{A}^1 \xrightarrow{\phi} \mathbb{A}^3$$

via $\phi(p) = (p, p^2, p^3)$. To find the equations of the image of this map, we form the graph

$$\Gamma = (p, \phi(p)) \subseteq \mathbb{A}^1 \times \mathbb{A}^3.$$

The equations of the graph are easy to write down, they are

$$x - t, y - t^2, z - t^3.$$

Of course, the image of ϕ is simply the projection of Γ onto $\mathbb{A}^3$, so to find the equations of it, we need to take a lex Gröbner basis with $t > x, y, z$.

```
i1 : R=ZZ/101[t,z,y,x, MonomialOrder=>Lex];

i2 : I=ideal(x-t,y-t^2,z-t^3)

                            2           3
o2 = ideal (- t + x, - t  + y, - t  + z)

o2 : Ideal of R

i3 : gb I

o3 = | y-x2 z-x3 t-x |

o3 : GroebnerBasis
```

In this example it is easy to see what the equations are directly. Try the following (by hand first, then check with your computer)

Exercise 4.4.6. Find equations for the image of the map

$$\mathbb{A}^2 \xrightarrow{\phi} \mathbb{A}^5$$

via $\phi(p, q) = (p, q, p^2, pq, q^2)$. $\diamond$

Supplemental reading: On Gröbner bases for ideals, Chapter 2 of [23] is simply unbeatable, for modules, see Chapter 15 of [28] or Chapter 5 of [24].

Other texts on Gröbner bases are Adams-Loustaunau [1] (which has many nice exercises - check out the one from Dave Bayer's thesis [7] on graph colorings) and Becker-Weispfenning [13]. For those interested in computational complexity, there are many issues which were not discussed here, but which are of real practical importance; two starting points are the papers of Mayr-Meyer [65] and Bayer-Stillman [12].

Chapter 5
Combinatorics, Topology and the Stanley–Reisner Ring

In the late 1800's the Italian mathematician Enrico Betti had the idea of modeling a surface by joining together a collection of triangles. This is sensible, since from a topological viewpoint there is no difference between the two-sphere S^2 and the boundary of a tetrahedron:

This simple idea turns out to be extremely fruitful, and generalizes naturally – we try to approximate a general topological space X using *simplices*, which are higher dimensional analogs of triangles. While this is not always possible, it works in many cases of interest. The union of the simplices is a *simplicial complex* Δ, which is a combinatorial approximation to X. From the data of Δ we can build a chain complex $C(\Delta)$. As you might expect, $C(\Delta)$ gives an algebraic encoding of information about X. In Chapter 2 we used free resolutions to understand graded modules, which showed how useful exact complexes are. One point of this chapter is that the failure of exactness can also be quite interesting. The homology (recall this measures failure of exactness) of $C(\Delta)$ is what captures nontrivial topological behavior.

There is also a beautiful connection between simplicial complexes and commutative rings. Given a simplicial complex Δ, we can build a commutative ring (the Stanley–Reisner ring), which knows everything about Δ! This means we can translate combinatorial problems into algebraic ones, which Stanley did to spectacular effect, as we'll see in Chapter 10. It also turns out that many essential algebraic theorems have pleasing combinatorial interpretations; the *primary decomposition* of the Stanley–Reisner ideal is purely combinatorial.

Key concepts: Simplex, simplicial complex, homology with coefficients, face vector, polytope, Stanley–Reisner ring.

5.1 Simplicial Complexes and Simplicial Homology

We begin with the definition of an *abstract n-simplex*:

Definition 5.1.1 (Abstract simplex). *Let V be a set of $n + 1$ elements. The n-simplex on V is the set of all subsets of V.*

If $V = \{v_1, v_2, v_3\}$, then the two-simplex on V is simply

$$\{\{v_1, v_2, v_3\}, \{v_1, v_2\}, \{v_1, v_3\}, \{v_2, v_3\}, \{v_1\}, \{v_2\}, \{v_3\}, \emptyset\}.$$

We now describe a geometric way to think of a simplex:

Definition 5.1.2 (Geometric simplex). *A set of $n + 1$ points $\{p_0, \ldots, p_n\} \subseteq \mathbb{R}^n$ is geometrically independent if the vectors $\{p_1 - p_0, p_2 - p_0, \ldots, p_n - p_0\}$ are linearly independent. The set of all convex combinations (see Exercise 1.1.1) of $n + 1$ geometrically independent points is called a geometric n-simplex.*

In $\mathbb{R}^n$ the origin and the points corresponding to the coordinate unit vectors are geometrically independent. The resulting n-dimensional simplex is called the *standard unit n-simplex*. We can associate to any abstract n-simplex the standard unit n-simplex, so for example we can think of an abstract two-simplex geometrically as a triangle and all of its faces, or a singleton set as a vertex. An *oriented simplex* is a simplex together with an ordering of the vertices; we define two orientations to be equivalent if they differ by an even permutation. We picture the orientation above as:

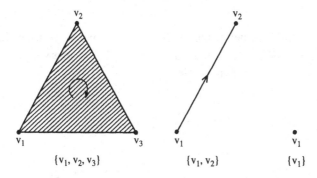

$\{v_1, v_2, v_3\}$ $\{v_1, v_2\}$ $\{v_1\}$

Notice that the edges are themselves simplices, so also have orientations.

Definition 5.1.3. *An abstract simplicial complex Δ on a vertex set V is a set of subsets of V such that*

- $\{v_i\} \in \Delta$ *if* $v_i \in V$
- *if* $\sigma \subseteq \tau \in \Delta$, *then* $\sigma \in \Delta$

The dimension of Δ is the dimension of the largest simplex it contains.

If an $(i+1)$-set τ is in Δ, then we call τ an i-face of Δ. The definition above just says that every vertex is a face of Δ, and that the relation "is a face of" is transitive. An abstract simplicial complex Δ gives a recipe for building a topological object $|\Delta|$, called the *geometric realization* of Δ: go to the simplex store, buy the number of standard unit n-simplices on your list, and attach them to each other as directed. See [71] for formalities on stretching and pasting these geometric objects so that they fit together.

Given a simplicial complex Δ and ring R, we define R-modules C_i as follows: the generators of C_i are the oriented i-simplices of Δ, and the relations are $\{v_{j_0}, \ldots, v_{j_i}\} = (-1)^{sgn(\sigma)}\{v_{j_{\sigma(0)}}, \ldots, v_{j_{\sigma(i)}}\}$ for $\sigma \in S_{i+1}$. Notice this captures the equivalence we defined earlier; C_i is free of rank equal to the number of i-faces of Δ. Technically, $c_i \in C_i$ is defined as a map from the oriented i-simplices to R which is zero except on finitely many simplices, and such that $c_i(\delta) = -c_i(\delta')$ if δ and δ' differ by an odd permutation. Since we'll always work with finite Δ, we don't need this level of formality. Recall from Chapter 2 that a chain complex is a sequence of objects and maps

$$\cdots F_{i+1} \xrightarrow{d_{i+1}} F_i \xrightarrow{d_i} F_{i-1} \xrightarrow{d_{i-1}} \cdots$$

with image $d_{i+1} \subseteq$ kernel d_i. One of the most important ways in which a chain complex arises is from a simplicial complex. The boundary of an oriented simplex $\{v_{i_0}, \ldots, v_{i_n}\}$ is defined via

$$\partial(\{v_{i_0}, \ldots, v_{i_n}\}) = \sum_{j=0}^{n} (-1)^j \{v_{i_0}, \ldots, \widehat{v_{i_j}}, \ldots, v_{i_n}\}.$$

For example, $\partial(\{v_1, v_2, v_3\}) = \{v_2, v_3\} - \{v_1, v_3\} + \{v_1, v_2\}$.

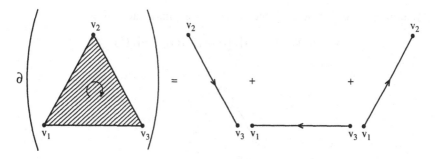

Exercise 5.1.4. Show that

$$\partial\partial = 0. \quad \diamond$$

Define a map

$$C_i \xrightarrow{\partial} C_{i-1},$$

by extending the ∂ map on oriented i-simplices via linearity. Elements of R are called coefficients, so we'll speak of computing homology with $\mathbb{Z}$, $\mathbb{Z}/2\mathbb{Z}$, or $\mathbb{Q}$ coefficients ($\mathbb{Z}$ is the default). By convention, $C_i = 0$ for $i > \dim \Delta$ and for $i < -1$. For $i = -1$, C_{-1} is a rank one free module, corresponding to the empty face. The homology of the resulting chain complex is called *reduced* homology, and is written $\widetilde{H}_i(\Delta)$. If we instead define $C_{-1} = 0$, then (check) we still have a chain complex, whose homology is written $H_i(\Delta)$. $\widetilde{H}_i$ and H_i only differ at $i = 0$, where

$$\text{rank } \widetilde{H}_0(\Delta) + 1 = \text{rank } H_0(\Delta).$$

Exercise 5.1.5. Prove that rank $H_0(\Delta)$ is the number of connected components of Δ. Thus,

$$\text{rank } \widetilde{H}_0(\Delta) = 0 \iff \Delta \text{ connected.}$$

Hint: if v_i and v_j lie in the same connected component, then there is a sequence of edges connecting them. $\diamond$

The rank of $H_i(\Delta)$ is called the ith Betti number of Δ. To avoid confusion, we will use lower case betti when referring to the ranks of the graded modules in a free resolution. The reason the Betti numbers are important is that they are topological invariants of $|\Delta|$. Munkres notes that proving this is "a reasonably arduous task"; a key ingredient is showing that the Betti numbers don't change under the operation of subdivision. The technical details can be found in Chapter 2 of [71]. Let's get our hands dirty with some examples.

Example 5.1.6. Let $R = \mathbb{Q}$ and consider the simplicial complex

$$\Delta = \{\{v_1, v_2\}, \{v_2, v_3\}, \{v_3, v_1\}, \{v_1\}, \{v_2\}, \{v_3\}, \emptyset\},$$

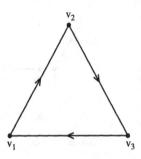

Then the vector spaces C_1 and C_0 are both three-dimensional. With respect to ordered bases $\{v_1, v_2\}$, $\{v_2, v_3\}$, $\{v_3, v_1\}$ and $\{v_1\}$, $\{v_2\}$, $\{v_3\}$, the differential ∂_1 is just -1 times the matrix which appeared in Exercise 2.3.1 (try writing down ∂_1 yourself before checking). In particular, $\dim H_1(\Delta) = 1$, $\dim H_0(\Delta) = 1$.

Exercise 5.1.7. Let Δ be a solid square, triangulated with a crossing diagonal, with triangles $\{\{v_1, v_2, v_3\}, \{v_1, v_4, v_2\}\}$ and edges $\{\{v_1, v_2\}, \{v_1, v_3\}, \{v_2, v_3\}, \{v_1, v_4\}, \{v_2, v_4\}\}$:

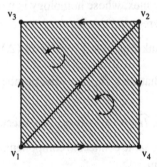

The chain complex with $\mathbb{Q}$ coefficients is:

$$0 \longrightarrow \mathbb{Q}^2 \xrightarrow{\partial_2} \mathbb{Q}^5 \xrightarrow{\partial_1} \mathbb{Q}^4 \longrightarrow 0,$$

and the boundary maps (with respect to bases ordered as above) are given by

$$\partial_2 = \begin{bmatrix} 1 & -1 \\ -1 & 0 \\ 1 & 0 \\ 0 & 1 \\ 0 & -1 \end{bmatrix}$$

$$\partial_1 = \begin{bmatrix} -1 & -1 & 0 & -1 & 0 \\ 1 & 0 & -1 & 0 & -1 \\ 0 & 1 & 1 & 0 & 0 \\ 0 & 0 & 0 & 1 & 1 \end{bmatrix}$$

1. For Δ as above, check that $H_2 = H_1 = 0$, and rank $H_0 = 1$.
2. Now remove $\{v_1, v_2, v_3\}$ and $\{v_1, v_4, v_2\}$ from Δ and compute homology. ◇

Macaulay 2 actually has a command called chainComplex. The command expects a series of matrices, which are the differentials of the complex. Since matrices have sources and targets, the modules in the complex are defined implicitly. Warning: when doing this with graded objects, care must be exercised to insure gradings agree. If we have a matrix M over R and a ring map $R \longrightarrow S$, then we can turn M into a matrix over S by *tensoring* (discussed in the next chapter). In Macaulay 2, the tensor command is **. We illustrate for the previous exercise:

```
i1 : R = QQ;

i2 : C=chainComplex(
matrix{{-1,-1,0,-1,0},{1,0,-1,0,-1},{0,1,1,0,0},
{0,0,0,1,1}}**R,
matrix{{1,-1},{-1,0},{1,0},{0,1},{0,-1}}**R)

         4        5       2
o2 = QQ  <-- QQ  <-- QQ

o2 : ChainComplex

i3 : C.dd
```

```
            4                                         5
o3 = 0 : QQ  <--------------------------- QQ    : 1
                 | -1 -1  0  -1 0 |
                 | 1   0 -1   0 -1 |
                 | 0   1  1   0  0 |
                 | 0   0  0   1  1 |
```

```
            5                                  2
     1 : QQ   <-------------- QQ   : 2
                  | 1   -1 |
                  | -1  0  |
                  | 1   0  |
                  | 0   1  |
                  | 0   -1 |

o3 : ChainComplexMap

i4 : C.dd_1 * C.dd_2

o4 = 0
--we check that the maps compose to 0
```

The command HH can be applied to a chain complex to compute the homology. We check one of the computations of the previous exercise:

```
i5 : rank HH_0 C

o5 = 1
```

Exercise 5.1.8. Identify the (outer) parallel edges of a rectangle as below:

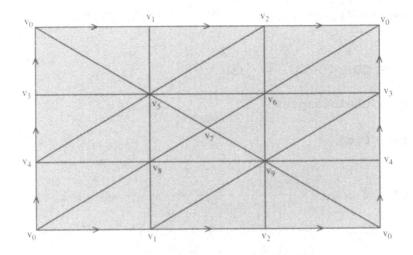

Then we obtain a triangulation of the torus:

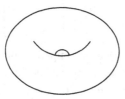

Orient the triangles counterclockwise (all the triangles are part of the complex Δ, so we omit shading), and the "interior" edges however you wish.

1. Compute homology with $\mathbb{Z}/2\mathbb{Z}$ coefficients, and with $\mathbb{Q}$ coefficients. Now change Δ so that on the right side of the rectangle the arrows point downward and vertices v_3 and v_4 are swapped. The geometric realization of this object is the *Klein bottle*. Try writing down ∂_2; you should get something just a bit different from the previous case. Again, compare $\mathbb{Q}$ and $\mathbb{Z}/2\mathbb{Z}$ coefficients. Check your answers in Chapter 1 of [71].

2. It would be quite a bit simpler if we could triangulate as below:

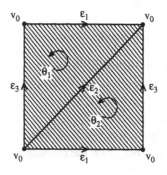

Of course, this would be cheating: the definition of a simplicial complex does not permit us to identify vertices in this way. Suspending our reservations for a moment and pretending all is well, we compute that $\partial(\theta_1) = -\epsilon_1 + \epsilon_2 - \epsilon_3$ and $\partial(\theta_2) = \epsilon_1 - \epsilon_2 + \epsilon_3$. This yields the chain complex (remember the vertices are all identified)

$$0 \longrightarrow R^2 \xrightarrow{\partial_2} R^3 \xrightarrow{\partial_1} R \longrightarrow 0,$$

where $\partial_1 = 0$ and

$$\partial_2 = \begin{bmatrix} -1 & 1 \\ 1 & -1 \\ -1 & 1 \end{bmatrix}$$

Compute the homology and compare your answers with the first part of the exercise. What just happened is that we used a *cellular* complex instead of a simplicial complex to compute homology. Notice (it is hard not to!) that the simplicial complex is quite a bit more complicated than the cellular complex. We won't say more about cellular complexes here, except to point out when you do see the definition of cellular homology, it looks complicated. As this exercise illustrates, there is a nice payoff. ◇

5.2 The Stanley–Reisner Ring

Given a simplicial complex Δ on n vertices, we associate to it a ring (the *Stanley-Reisner ring*)

$$k[\Delta] = k[v_1, \ldots, v_n]/I_\Delta,$$

where the v_i are the vertices of Δ and I_Δ is the ideal generated by monomials corresponding to *nonfaces* of Δ.

Example 5.2.1. Examples of Stanley–Reisner rings

1. If Δ is an n-simplex, then $k[\Delta] = k[v_1, \ldots, v_{n+1}]$
2. Take Δ as in Exercise 5.1.7.1. Then every nonface includes $v_3 v_4$, so $k[\Delta] = k[v_1, \ldots, v_4]/\langle v_3 v_4 \rangle$
3. Take Δ as in Exercise 5.1.7.2. Then $I_\Delta = \langle v_3 v_4, v_1 v_2 v_3, v_1 v_2 v_4 \rangle$.
4. If Δ is the boundary of an octahedron, then I_Δ is generated by the three edges $v_i v_j$ which do not lie on the boundary.

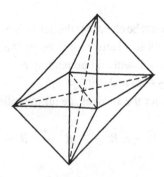

Exercise 5.2.2. Prove that if $\{v_{i_1}, \ldots, v_{i_k}\}$ is not a face of Δ, then no superset of $\{v_{i_1}, \ldots, v_{i_k}\}$ is a face of Δ. ◇

The Stanley–Reisner ring of Δ encodes lots of useful topological and combinatorial information about Δ. One really important combinatorial invariant of a simplicial complex is the f-vector. The letter f stands for face: f_i is the number of i-dimensional faces of Δ; by convention there is one empty face, so the f-vector starts with a 1 in position -1. For example, the f-vector of a line segment connecting two vertices would be $(1, 2, 1)$.

One much studied class of simplicial complexes are simplicial polytopes. A *polytope* P on n vertices is the set of all convex combinations (convex hull) of n points in some Euclidean space (assuming no vertex is a convex combination of the other vertices). We say that P is d-dimensional if $\mathbb{R}^d$ is the smallest Euclidean space containing P (so P is topologically a d-ball) and that P is *simplicial* if all the faces of the boundary ∂P are simplices; by convention "the f-vector of P" *means* $f(\partial P)$. An octahedron is a simplicial polytope, whereas the cube and pyramid with a rectangular base are non-simplicial. If we fix a number of vertices, what f-vectors can arise as the f-vector of a simplicial polytope? What is the (componentwise) biggest f-vector that can occur? Should a biggest f-vector even exist? The answers use the Stanley–Reisner ring! We'll return to this question in Chapter 10.

Exercise 5.2.3. For a three-dimensional simplicial polytope, prove that the following two relations hold:

$$f_0 - f_1 + f_2 = 2, \text{ and } 3f_2 = 2f_1.$$

So if we know f_0, then we know the f-vector. Hint: for the first relation (known as Euler's relation) get really close to the polytope and peer through one of the faces – you see a planar graph (technically, this is called a *Schlegel* diagram), so just run an induction (don't forget to include the face you peered through!). For the second relation, count edge/triangle pairs two different ways. A really cool proof of Euler's relation (using the Poincaré-Hopf index theorem!) can be found in [41]. ◇

It turns out that it is often useful to rewrite the f-vector of a $d - 1$ complex in terms of the h-vector:

$$h_j = \sum_{i=0}^{j}(-1)^{j-i}\binom{d-i}{j-i}f_{i-1}, \text{ and } f_j = \sum_{i=0}^{j+1}\binom{d-i}{j+1-i}h_i.$$

This seems confusing, but there is a nice trick of Stanley which makes it quick to compute by hand (we illustrate for Δ two-dimensional). Draw a triangle, with the f-vector down the right side, and ones on the left:

$$
\begin{array}{cccc}
& & 1 & \\
& 1 & & f_0 \\
& 1 & & f_1 \\
1 & & & f_2 \\
1 & & &
\end{array}
$$

At an interior position p in the triangle we define $p = $ (value immediately to the Northeast of p) − (value immediately to the Northwest of p). For example, in the third row, we put $f_0 - 1$ in the middle position. For the boundary of the octahedron we have:

$$
\begin{array}{ccccccc}
& & & 1 & & & \\
& & 1 & & 6 & & \\
& 1 & & 5 & & 12 & \\
1 & & 4 & & 7 & & 8 \\
1 & & 3 & & 3 & & 1
\end{array}
$$

The h-vector appears in the bottom row. As usual, we might well ask why the h-vector is useful. Let's first find a free resolution of the Stanley–Reisner ring; label the picture on the previous page and do so; you should be able to do it faster by hand than by computer (don't peek at the solution below till you've tried it!).

$$
0 \longrightarrow R(-6) \longrightarrow R(-4)^3 \longrightarrow R(-2)^3 \longrightarrow R \longrightarrow R/I_\Delta \longrightarrow 0.
$$

If we compute the Hilbert series for R/I_Δ, we obtain

$$
\frac{1 - 3t^2 + 3t^4 - t^6}{(1-t)^6} = \frac{1 + 3t + 3t^2 + t^3}{(1-t)^3}.
$$

So h_i is the coefficient of t^i in the numerator of the simplified Hilbert series!

```
i6 : R=ZZ/101[v_1..v_6];

i7 : i = matrix{{v_1*v_4, v_2*v_5,v_3*v_6}}

o7 = {0} | v_1v_4 v_2v_5 v_3v_6 |

                1        3
o7 : Matrix R    <--- R

i8 : res coker i
```

```
          1       3       3       1
o8 = R   <--  R   <--  R   <--  R

          0       1       2       3

o8 : ChainComplex

i9 : o8.dd
                        1
o9 = -1 : 0 <-----  R   : 0
                 0


        1                                                 3
0 : R   <--------------------------------------- R   : 1
            {0} | v_1v_4 v_2v_5 v_3v_6 |


        3                                                 3
1 : R   <--------------------------------------- R   : 2
            {2} | -v_2v_5 -v_3v_6  0          |
            {2} | v_1v_4   0       -v_3v_6 |
            {2} | 0        v_1v_4  v_2v_5  |


        3                         1
2 : R   <------------------- R   : 3
            {4} | v_3v_6  |
            {4} | -v_2v_5 |
            {4} | v_1v_4  |

o9 : ChainComplexMap

i10 : poincare coker i

              6       4       2
o10 = - $T  + 3$T  - 3$T  + 1

o11 : ZZ[ZZ^1]

i11 : factor o10

                  3       3
o11 = ($T - 1) ($T + 1) (- 1)
```

Exercise 5.2.4. Prove that for a simplicial three-polytope, the h-vector is the numerator of the Hilbert series. Use the relation between the f and h-vectors and the lemma below. $\diamond$

Lemma 5.2.5. *The Hilbert polynomial of the Stanley–Reisner ring of a $(d-1)$-dimensional simplicial complex Δ is:*

$$HP(R/I_\Delta, i) = \sum_{j=0}^{d-1} f_j \binom{i-1}{j}.$$

Proof. We count monomials in the Stanley–Reisner ring in terms of their support (the support of a monomial m is just the set of indices k such that v_k divides m nontrivially, e.g., the support of $v_1^2 v_4^3$ is $\{1, 4\}$). The point is that a monomial which is not in I_Δ must be supported on one of the faces of Δ. So, say we want to count the degree i monomials. If a monomial is supported on a vertex v_k, and is of degree i, then it must just be v_k^i, hence there are f_0 monomials supported only on vertices. For an edge $v_k v_l$, the monomial must be of the form $v_k^a v_l^{i-a}$ and both a and $i-a$ must be bigger than one, i.e. we have $i-2$ degrees to distribute into two boxes, which is $\binom{i-2+1}{1}$ (in general, think of adding spaces where we will place dividers), so we have $f_1 \binom{i-1}{1}$ monomials supported only on edges. Continue in this way up to the maximal faces, done. $\square$

There are other reasons why the h-vector is a nice object. For a simplicial polytope, we saw in Exercise 5.2.3 that there are relations between the f_i. When expressed in terms of the h-vector these relations are very simple.

Exercise 5.2.6. Use Exercise 5.2.3 and Stanley's trick to prove that the h-vector of a simplicial three-polytope is symmetric. $\diamond$

The relations $h_i = h_{d-i}$ hold for any simplicial d-polytope, and are called the *Dehn–Sommerville* relations. The proof of this is quite beautiful and elementary, and can be found in Chapter 8 of [100]. These relations also have another interpretation. A simplicial polytope gives rise to an object called a toric variety. A fundamental result in homology theory is *Poincaré duality*, which says that the vector of Betti numbers of a connected, compact, oriented n-manifold is symmetric. The complex toric variety $X(\Delta)$ associated to a simplicial polytope Δ is compact and close enough to being an oriented manifold that Poincaré duality holds; the h_i of Δ are also the even Betti numbers H_{2i} of $X(\Delta)$!

5.3 Associated Primes and Primary Decomposition

Stanley–Reisner rings have particularly pretty primary decompositions:

Example 5.3.1. If Δ is the complex of Exercise 5.1.7.2, then

$$I_\Delta = \langle v_3 v_4, v_1 v_2 v_3, v_1 v_2 v_4 \rangle,$$

and we can write

$$I_\Delta = \langle v_1, v_3 \rangle \cap \langle v_1, v_4 \rangle \cap \langle v_2, v_3 \rangle \cap \langle v_2, v_4 \rangle \cap \langle v_3, v_4 \rangle.$$

Check this using the Macaulay 2 command intersect – your output should look like:

```
i13 : intersect(ideal(v_1,v_3),ideal(v_1,v_4),
               ideal(v_2,v_3),ideal(v_2,v_4),
               ideal(v_3,v_4))

o13 = ideal (v v , v v v , v v v )
             3 4   1 2 4   1 2 3
```

Try some more examples for Stanley–Reisner rings, and see if you can come up with a nice description of the primary decomposition. Hint: first, see if you can find some relation involving *cofaces* (a coface is a set of vertices whose complement is a face). The following lemma from Chapter 1 may also be useful:

Lemma 5.3.2. *For R Noetherian, if $I = \cap Q_i$, with Q_i primary to P_i, then the P_i are the prime ideals occurring in $\{\sqrt{I : f} \mid f \in R\}$.*

Don't read further until you have run some experiments and come up with a conjecture (or better, proof!)

Theorem 5.3.3. *Let Δ be a simplicial complex. Then*

$$I_\Delta = \bigcap_{v_{i_1} \cdots v_{i_k} \text{ a minimal coface}} \langle v_{i_1}, \ldots, v_{i_k} \rangle.$$

Proof. First, if $\sigma = v_{j_1} \cdots v_{j_n}$ is a maximal face, then $v_{j_1} \cdots v_{j_n} v_k$ is a nonface for any $k \notin \{j_1, \ldots, j_n\}$, i.e. for σ as above

$$I_\Delta : v_{j_1} \cdots v_{j_n} = \langle v_k | k \notin \{j_1, \ldots, j_n\}\rangle.$$

This is prime by Exercise 4.2.6, so by Lemma 5.3.2 we have

$$I_\Delta \subseteq \bigcap_{v_{i_1} \cdots v_{i_k} \text{ a minimal coface}} \langle v_{i_1}, \ldots, v_{i_k}\rangle.$$

On the other hand, if a monomial is in the ideal on the right, then by the way we built it, it is not supported on any maximal face (it has a variable in every coface), hence not supported on any face, so lies in I_Δ. □

Example 5.3.4. Consider the simplicial complex consisting of all the edges of a tetrahedron, and a single triangle:

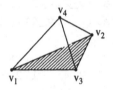

$$\Delta = \{\{v_1, v_2, v_3\}, \{v_1, v_2\}, \{v_1, v_3\}, \{v_2, v_3\}, \{v_1, v_4\}, \{v_2, v_4\}, \{v_3, v_4\},$$
$$\{v_1\}, \{v_2\}, \{v_3\}, \{v_4\}, \emptyset\}.$$

Then the maximal faces are

$$\{\{v_1, v_2, v_3\}, \{v_1, v_4\}, \{v_2, v_4\}, \{v_3, v_4\}\},$$

so the minimal cofaces are

$$\{\{v_4\}, \{v_2, v_3\}, \{v_1, v_3\}, \{v_1, v_2\}\}.$$

Thus, the primary decomposition of I_Δ is

$$I_\Delta = \langle v_4 \rangle \cap \langle v_2, v_3 \rangle \cap \langle v_1, v_3 \rangle \cap \langle v_1, v_2 \rangle =$$
$$\langle v_4 \rangle \cap \langle v_1 v_2, v_1 v_3, v_2 v_3 \rangle = \langle v_1 v_2 v_4, v_1 v_3 v_4, v_2 v_3 v_4 \rangle,$$

as expected.

```
i14 :   decompose(ideal(v_1*v_2*v_4, v_1*v_3*v_4,
                        v_2*v_3*v_4))
```

$$o14 = \{\text{ideal } v_4 , \text{ ideal } (v_1 , v_2), \text{ ideal } (v_3 , v_2),$$
$$\text{ideal } (v_3 , v_1)\}$$

Exercise 5.3.5. Write a Macaulay 2 script which takes as input a list of the maximal simplices of a simplicial complex Δ, and which returns as output the Stanley–Reisner ideal and the chain complex $C(\Delta)$. $\diamond$

Supplemental reading: For simplicial complexes and the Stanley–Reisner ring, see Stanley [88], for polytopes, see Ziegler [100], and for homology, see Munkres [71]. For associated primes and primary decomposition, Chapter 4 of [3] or Chapter 3 of [28] are good, and also section 4.7 of [23]. There is quite a bit of current research activity in this area, especially related to understanding the free resolution of R/I_Δ. This started in [55], where Hochster gave a beautiful combinatorial interpretation for the betti numbers. Recently Eagon and Reiner [27] showed that the betti numbers appear on a single row (of the Macaulay 2 betti diagram) iff the *Alexander dual* of Δ is arithmetically Cohen-Macaulay (the Cohen-Macaulay property is discussed in Chapter 10, see Miller [68] for Alexander duality). For other recent progress see papers [4], [5] of Aramova, Herzog, and Hibi, and for work on monomial ideals which are not square-free see Bayer-Charalambous-Popescu [8] and Bayer-Peeva-Sturmfels [10]. Work on simplicial polytopes (with fixed h-vector) having maximal betti numbers appears in [67]. The books [92], [93] of Sturmfels describe many other connections between algebra and combinatorics not discussed here, and are also filled with interesting open problems.

Chapter 6

Functors: Localization, Hom, and Tensor

A really important idea in mathematics is to try to use techniques from one area to solve problems in another area. Of course different areas speak different languages, so we need a translator. Such a translator is called a *functor*, and in this lecture we'll study the three most important functors in algebra: localization, hom, and tensor. When we translate problems, there are basically two things to take into account: what the objects are, and what the maps between objects are. Formally, a *category* consists of objects and morphisms between the objects, and a functor is a rule which sends objects/morphisms in a category C to objects/morphisms in another category D. If we lose too much information, then the translation is useless, so we very reasonably require that a functor preserve identity morphisms and compositions. If M_1 and M_2 are objects in C, and f is a morphism

$$M_1 \xrightarrow{f} M_2,$$

then a functor F from C to D is *covariant* if

$$F(M_1) \xrightarrow{F(f)} F(M_2),$$

and *contravariant* if the direction of the map gets switched:

$$F(M_1) \xleftarrow{F(f)} F(M_2).$$

For example, let R be a ring and C the category of R-modules and R-module homomorphisms. If N is a fixed R-module, $Hom_R(\bullet, N)$ is a contravariant functor from C to C, whereas $Hom_R(N, \bullet)$ is covariant. Since functors deal with objects and morphisms, it is natural to ask how a functor transforms a short exact sequence. The three functors we study in this section provide three important examples of the possibilities.

Key Concepts: Functor, left exact, right exact, covariant, contravariant, localization, Hom, $\otimes$.

6.1 Localization

One way to simplify an algebraic object (be it vector space, group, ring, or module) is to quotient out by a subobject. If we're interested in an object M, and we have a subobject M', then we can (via an exact sequence) hope to understand M by understanding M' and M/M'. In a sense, when we quotient by M', we erase it. *Localization* is a way of simplifying an object, but instead of zeroing out some set, we make the objects in the set into units. Here is the formal definition:

Definition 6.1.1. *Let R be a ring, and S a multiplicatively closed set containing* 1. *Define an equivalence relation on $\{\frac{a}{b} | a \in R, b \in S\}$ via*

$$\frac{a}{b} \sim \frac{c}{d} \text{ if } (ad - bc)u = 0 \text{ for some } u \in S.$$

Then the localization of R at S is

$$R_S = \left\{ \frac{a}{b} | a \in R, b \in S \right\} / \sim .$$

Example 6.1.2. Let $R = \mathbb{Z}$, and let S consist of all nonzero elements of R. Clearly $R_S = \mathbb{Q}$, so R_S is in fact a ring, with a very natural ring structure.

There are two very common cases of sets which play the role of S in this construction. The first is when S is the complement of a prime ideal, and the second is when $S = \{1, f, f^2, \ldots\}$ for some $f \in R$; it is easy to see these sets are multiplicatively closed. In the first case, if the prime ideal is P, we often write R_P for R_S, notice that by construction R_P has a unique maximal ideal so R_P is a local ring. For an R-module M, there is an obvious way to construct M_S; in particular, localization is a functor from (R-modules, R-hom's) to (R_S-modules, R_S-hom's). A very nice property of localization is that it is an *exact functor*:

Theorem 6.1.3. *Localization preserves exact sequences.*

Proof. First, suppose we have a map of R-modules $M \xrightarrow{\phi} M'$. Since ϕ is R-linear, this gives us a map $M_S \xrightarrow{\phi_S} M'_S$ via $\phi_S(\frac{m}{s}) = \frac{\phi(m)}{s}$. Now suppose we have an exact sequence

$$0 \longrightarrow M' \xrightarrow{\phi} M \xrightarrow{\psi} M'' \longrightarrow 0.$$

It is easy to check that

$$\psi_S \phi_S = 0,$$

so suppose that $\frac{m}{s'} \in \ker \psi_S$, so that $\frac{\psi(m)}{s'} \sim \frac{0}{s''}$, hence there is an $s \in S$ such that $s\psi(m) = 0$ in R. But $s\psi(m) = \psi(sm)$ so $sm \in \ker \psi = \operatorname{im} \phi$, i.e., $sm = \phi(n)$. Thus, we have $m = \frac{\phi(n)}{s}$ and

$$\frac{m}{s'} = \frac{\phi(n)}{ss'}. \qquad \square$$

Exercise 6.1.4. Let M be a finitely generated R-module, and S a multiplicatively closed set. Show that $M_S = 0$ iff $\exists s \in S$ such that $s \cdot M = 0$. ◇

The *support* of an R-module M consists of those primes P such that $M_P \neq 0$.

Exercise 6.1.5. The *annihilator* of an R-module M is $\{r \in R | r \cdot M = 0\}$. If M is finitely generated, show that P is in the support of M iff $ann(M) \subseteq P$. ◇

Example 6.1.6. Let $\mathcal{A}$ be an arrangement of d distinct lines in $\mathbb{P}^2_\mathbb{Q}$. For each line λ_i of $\mathcal{A}$, fix $l_i \in R = \mathbb{Q}[x, y, z]$ a nonzero linear form vanishing on λ_i, put $f = \prod_{i=1}^d l_i$, and let J_f be the Jacobian ideal of f

$$J_f = \left\langle \frac{\partial f}{\partial x}, \frac{\partial f}{\partial y}, \frac{\partial f}{\partial z} \right\rangle.$$

It follows from Example A.3.2 that $p \in V(J_f)$ iff p is a point where the lines of $\mathcal{A}$ meet. The ideal J_f is not generally radical, so if we want to compute the Hilbert polynomial of R/J_f, we must do more than just count the points of intersection. Recall from Chapter 3 that a complete intersection ideal is an ideal generated by a regular sequence; a homogeneous ideal I is a *local complete intersection* if I_P is a complete intersection in R_P for any prime ideal P corresponding to a point of $V(I)$.

Exercise 6.1.7. Prove that the Jacobian ideal of a line arrangement is a local complete intersection. Hint: let p be a point where some of the lines meet; we may assume p is the point $(0:0:1)$ so $P = \langle x, y \rangle$. Write $f = L_p L_0$, with L_p the product of those linear forms passing through $(0:0:1)$ and L_0 the product of the remaining forms. Differentiate using the product rule (enter localization: after differentiating, you can *considerably* simplify the result since in the local ring R_P any polynomial involving a monomial z^n is invertible). Finally, apply Euler's relation: if $F \in \mathbb{Q}[x_1, \dots, x_n]$ is homogeneous of degree α, then

$$\alpha F = \sum_{i=1}^n x_i \frac{\partial F}{\partial x_i}. \qquad ◇$$

A consequence of the preceding exercise is that the P-primary component Q in the primary decomposition of J_f is given by $\langle \frac{\partial L_p}{\partial x}, \frac{\partial L_p}{\partial y} \rangle$; since L_p has no repeat factors, this is a complete intersection, hence the Hilbert polynomial $HP(R/Q, i) = (\deg L_p - 1)^2$. If we write the primary decomposition of J_f as $\bigcap_j Q_j \cap n$, where $\sqrt{n} = \langle x, y, z \rangle$, then a slight modification of Exercise 2.3.5 shows that

$$HP(R/J_f, i) = \sum_j HP(R/Q_j, i).$$

For a point p_j where t_j lines meet, define $\mu(p_j) = t_j - 1$. Then we have just shown that the Hilbert polynomial of R/J_f is the sum over all points of $\mu(p_j)^2$. Here is an example: take the lines $\{x = 0\}$, $\{y = 0\}$, $\{z = 0\}$, $\{x + z = 0\}$, $\{y + z = 0\}$. So if we draw the lines in $\mathbb{P}^2$ with $\{z = 0\}$ as the line at infinity, this is:

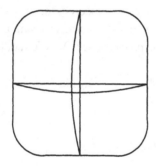

As we see from the picture, there are four points where two lines meet, and two points where three lines meet (remember, antipodal points are identified!), so we expect

$$HP(R/J_f, i) = 1 + 1 + 1 + 1 + 2^2 + 2^2 = 12.$$

We check with Macaulay 2:

```
i1 : R=QQ[x,y,z];

i2 : f = ideal(x*y*z*(x+z)*(y+z));

o2 : Ideal of R

i3 : J = jacobian f
```

```
o3 = {1} | 2xy2z+2xyz2+y2z2+yz3   |
     {1} | 2x2yz+x2z2+2xyz2+xz3   |
     {1} | x2y2+2x2yz+2xy2z+3xyz2 |

              3        1
o3 : Matrix R   <--- R

i4 : hilbertPolynomial coker transpose J

o4 = 12*P
        0
```

We close by mentioning a famous open question on line arrangements: do the combinatorics of the set of lines determine when the free resolution of R/J_f has length two? Check that the example above does have a free resolution of length two (by the Hilbert syzygy theorem the resolution has length at most three), and that moving the lines (holding the combinatorics fixed) does not alter this. For more on the problem, see Orlik-Terao, [75].

Exercise 6.1.8. Which of the following ideals of $\mathbb{Q}[x, y, z]$ are local complete intersections?

1. $\langle xy, xz, yz \rangle$
2. $\langle x^2z + xy^2, xyz + 2y^3, y^2z + x^3 \rangle$
3. $\langle x^2, xy, y^2 \rangle$ $\diamond$

6.2 The Hom Functor

Suppose we have a homomorphism of R-modules

$$M_1 \xrightarrow{\phi} M_2,$$

and

$$\alpha \in Hom_R(M_2, N).$$

Then we can cook up an element of $Hom_R(M_1, N)$, simply by composing α with ϕ.

$$M_1 \xrightarrow{\phi} M_2$$
$$\downarrow \alpha$$
$$N$$

In other words, we have a map:

$$Hom_R(M_2, N) \longrightarrow Hom_R(M_1, N),$$

where $\alpha \longrightarrow \alpha \cdot \phi$. In particular, $Hom_R(\bullet, N)$ is a contravariant functor from the category of R-modules and R-module homomorphisms to itself. Notice that if we instead fix the first "coordinate" we also obtain a functor $Hom_R(N, \bullet)$, which (check!) is covariant.

Exercise 6.2.1. Show that a short exact sequence of R-modules

$$0 \longrightarrow A_1 \xrightarrow{a_1} A_2 \xrightarrow{a_2} A_3 \longrightarrow 0$$

gives rise to a left exact sequence:

$$0 \longrightarrow Hom_R(A_3, N) \longrightarrow Hom_R(A_2, N) \longrightarrow Hom_R(A_1, N) \quad \Diamond$$

Here is an easy example showing that the rightmost map may not be surjective. In other words, unlike localization, which is an exact functor, $Hom_R(\bullet, N)$ is only a *left exact functor*.

Example 6.2.2. Let $1 \neq p \in \mathbb{Z}$, and consider the following exact sequence of $\mathbb{Z}$-modules:

$$0 \longrightarrow \mathbb{Z} \xrightarrow{\cdot p} \mathbb{Z} \longrightarrow \mathbb{Z}/p\mathbb{Z} \longrightarrow 0.$$

Let's apply $Hom_{\mathbb{Z}}(\bullet, \mathbb{Z})$ to this sequence. Remark: if F is a free $\mathbb{Z}$-module, then $Hom_{\mathbb{Z}}(F, \mathbb{Z})$ is isomorphic to F, because the only elements of $Hom_{\mathbb{Z}}(\mathbb{Z}, \mathbb{Z})$ are multiplication by an element of $\mathbb{Z}$. We have

$$0 \longrightarrow Hom_{\mathbb{Z}}(\mathbb{Z}/p\mathbb{Z}, \mathbb{Z}) \longrightarrow Hom_{\mathbb{Z}}(\mathbb{Z}, \mathbb{Z}) \longrightarrow Hom_{\mathbb{Z}}(\mathbb{Z}, \mathbb{Z}),$$

where the rightmost map is $m \to pm$. But this cannot be surjective; in particular the identity map is not of this form.

If G is an Abelian group and

$$G \simeq \mathbb{Z}^n \oplus G_{torsion},$$

then

$$Hom_{\mathbb{Z}}(G, \mathbb{Z}) \simeq \mathbb{Z}^n.$$

In other words, $Hom_{\mathbb{Z}}(\bullet, \mathbb{Z})$ kills torsion. In light of this, let's reexamine the sequence above: obviously every element of $\mathbb{Z}/p\mathbb{Z}$ is torsion, so

$$Hom_{\mathbb{Z}}(\mathbb{Z}/p\mathbb{Z}, \mathbb{Z}) \simeq 0,$$

and in the exact sequence of *Hom* modules, we obtain:

$$0 \longrightarrow Hom_{\mathbb{Z}}(\mathbb{Z}/p\mathbb{Z}, \mathbb{Z}) \simeq 0 \longrightarrow Hom_{\mathbb{Z}}(\mathbb{Z}, \mathbb{Z}) \simeq \mathbb{Z} \xrightarrow{\cdot p} Hom_{\mathbb{Z}}(\mathbb{Z}, \mathbb{Z}) \simeq \mathbb{Z}$$
$$\longrightarrow coker(\cdot p) \simeq \mathbb{Z}/p\mathbb{Z} \longrightarrow 0.$$

So we have another way to see that the sequence

$$0 \longrightarrow Hom_{\mathbb{Z}}(\mathbb{Z}/p\mathbb{Z}, \mathbb{Z}) \longrightarrow Hom_{\mathbb{Z}}(\mathbb{Z}, \mathbb{Z}) \longrightarrow Hom_{\mathbb{Z}}(\mathbb{Z}, \mathbb{Z}),$$

is not exact: the rightmost map has a cokernel, in particular it is not surjective. Who cares? What good is *Hom*? Well, let's flashback to the last chapter: one place where applying the *Hom* functor has nice consequences is in the case of simplicial homology. Recall that $C_i(\Delta)$ is a free R-module (R the coefficient ring) with basis the oriented i-simplices of Δ. We define a new chain complex (with arrows going in the opposite way) as

$$C^i(\Delta) = Hom_R(C_i(\Delta), R).$$

Exercise 6.2.3. Verify that the $C^i(\Delta)$ form a chain complex. The differential, of course, comes from the beginning of this section. ◇

The really neat thing is that the homology of the new chain complex (which is called simplicial *cohomology*) has important additional structure, which is not shared by the homology of the original chain complex. In particular, the simplicial cohomology is a ring!

Definition 6.2.4. *The multiplicative structure (called cup product) on simplicial cohomology is induced by a map on the cochains:*

$$C^i(\Delta) \times C^j(\Delta) \xrightarrow{\cup} C^{i+j}(\Delta);$$

it is defined by saying how a pair $(c_i, c_j) \in C^i(\Delta) \times C^j(\Delta)$ acts on an element of $C_{i+j}(\Delta)$: if $(v_0, \ldots v_{i+j}) \in C_{i+j}(\Delta)$, then

$$(c_i, c_j)(v_0, \ldots v_{i+j}) = c_i(v_0, \ldots, v_i) \cdot c_j(v_i, \ldots, v_{i+j}).$$

Exercise 6.2.5. Compute the simplicial cohomology ring of the two-sphere S^2 (hint: the simplicial complex with maximal faces $\{1, 2, 3\}$, $\{1, 2, 4\}$, $\{1, 3, 4\}$, $\{2, 3, 4\}$ corresponds to a hollow tetrahedron Δ whose geometric realization $|\Delta| \simeq S^2$). Next, determine the multiplicative structure of the cohomology ring of the torus. If you get stuck, you can find the answer in [71]. ◇

Of course, the case of most interest to us is the situation where R is a polynomial ring. To give a homomorphism between finitely generated modules M_1 and M_2, we first need to say where the generators of M_1 go. Of course, a homomorphism of R-modules must preserve relations on the elements of M_1. In sum, given presentations

$$R^{a_1} \xrightarrow{\alpha} R^{a_0} \longrightarrow M_1 \longrightarrow 0,$$

and

$$R^{b_1} \xrightarrow{\beta} R^{b_0} \longrightarrow M_2 \longrightarrow 0,$$

an element of $Hom_R(M_1, M_2)$ may be thought of as a map $R^{a_0} \xrightarrow{\gamma} R^{b_0}$; to encode the fact that relations in M_1 must go to zero we require that if c_α is in the image of α, then $\gamma(c_\alpha)$ is in the image of β. So the image of the composite map

$$R^{a_1} \xrightarrow{\gamma \cdot \alpha} R^{b_0}$$

is contained in the image of β.

Example 6.2.6. Here is how to have Macaulay 2 find the *entire* module of homomorphisms between two modules.

```
i1 : R=ZZ/101[x,y,z];

i2 : M = coker matrix {{x,y},{y,z}};

i3 : N = coker matrix {{y}};

i4 : Hom(M,N)

o4 = subquotient (0, | 0 y |)
                      | y 0 |

                                     2
o4 : R-module, subquotient of R

i5 : prune o4

o5 = 0
```

For matrices f and g with the same target, subquotient(f,g) is a module representing the image of f in the cokernel of g. This module may not be minimally presented; the prune command finds a minimal presentation. So $Hom(M, N)$ is zero. To check this, let R^2 have basis ϵ_1, ϵ_2. If γ is an element of $Hom(M, N)$, then in M, $x\epsilon_1 + y\epsilon_2$ and $y\epsilon_1 + z\epsilon_2$ are both zero, so that $0 = \gamma(x\epsilon_1 + y\epsilon_2) = x\gamma(\epsilon_1) + y\gamma(\epsilon_2)$. Now y is zero in N, so we get $0 = x\gamma(\epsilon_1)$. It is obvious that x is a nonzerodivisor on N, hence $\gamma(\epsilon_1) = 0$. Argue similarly for $\gamma(\epsilon_2)$ to see that γ must be the zero map.

Exercise 6.2.7. In the category of graded $R = k[x_1, \ldots, x_n]$-modules, show that $Hom_R(R(-a), R) \simeq R(a)$ (see Example 2.3.2). $\diamond$

6.3 Tensor Product

We begin by quickly reviewing tensor product. Suppose we have two R-modules M and N.

Definition 6.3.1. *Let A be the free R-module generated by $\{(m, n)|m \in M, n \in N\}$, and let B be the submodule of A generated by*

$$(m_1 + m_2, n) - (m_1, n) - (m_2, n)$$
$$(m, n_1 + n_2) - (m, n_1) - (m, n_2)$$
$$(rm, n) - r(m, n)$$
$$(m, rn) - r(m, n).$$

The tensor product is the R-module:

$$M \otimes_R N := A/B.$$

We write $m \otimes n$ to denote the class (m, n).

This seems like a nasty definition, but once you get used to it, tensor product is a natural thing. The relations $(rm, n) \sim r(m, n)$ and $(m, rn) \sim r(m, n)$ say that tensor product is R-linear, and turn out to be very useful.

Example 6.3.2. What is

$$\mathbb{Z}/3\mathbb{Z} \otimes_{\mathbb{Z}} \mathbb{Z}/2\mathbb{Z}?$$

If we have a zero in either coordinate, then we can pull zero out, so the only potential nonzero elements of this $\mathbb{Z}$-module are

$$2 \otimes 1, 1 \otimes 1.$$

But $2 \otimes 1 = 2(1 \otimes 1) = 1 \otimes 2 = 1 \otimes 0 = 0$, so both these elements are zero. Hence the $\mathbb{Z}$-module $\mathbb{Z}/3\mathbb{Z} \otimes_{\mathbb{Z}} \mathbb{Z}/2\mathbb{Z}$ is zero.

Exercise 6.3.3. Generalize the above: for $a, b \in \mathbb{Z}$, show that

$$\mathbb{Z}/a\mathbb{Z} \otimes_{\mathbb{Z}} \mathbb{Z}/b\mathbb{Z} \simeq \mathbb{Z}/GCD(a,b)\mathbb{Z}. \quad \diamond$$

If M and N are R-modules, then a map

$$M \times N \xrightarrow{f} P$$

is *bilinear* if $f(rm_1 + m_2, n) = rf(m_1, n) + f(m_2, n)$, and similarly in the second coordinate. Tensor product converts R–bilinear maps into R–linear maps, and possesses a *universal mapping property*: given a bilinear map f, there is a unique R–linear map $M \otimes_R N \longrightarrow P$ making the following diagram commute:

$$\begin{array}{ccc} M \times N & \xrightarrow{f} & P \\ \downarrow & \nearrow & \\ M \otimes N & & \end{array}$$

Exercise 6.3.4. Prove the universal mapping property of tensor product. $\diamond$

The most common application of tensor product is *extension of scalars*–we first set the stage. If $A \xrightarrow{f} B$ is a ring homomorphism, and M is a B-module, then we can make M an A-module in the obvious way: $a \in A$ acts on $m \in M$ via

$$a \cdot m = f(a) \cdot m.$$

This is usually called restriction of scalars. What if we have an A-module? How can we make it a B-module? Well, B is itself an A-module via $a \cdot b = f(a) \cdot b$, so we have a pair of A-modules. Tensor them! The module $B \otimes_A M$ is an A-module, but also a B-module. Extension of scalars is a common and important construction. Caution: in this book, R is always a commutative ring; care must be exercised when using tensor product over a noncommutative ring.

Exercise 6.3.5. Let $k \hookrightarrow k[x]$ play the role of $A \xrightarrow{f} B$, and let M be the k-module k^2. Describe the $k[x]$-module $k[x] \otimes_k k^2$. $\diamond$

As in the last section, suppose we have a short exact sequence of R-modules:

$$0 \longrightarrow A_1 \longrightarrow A_2 \longrightarrow A_3 \longrightarrow 0,$$

and let M be some other R-module. When we apply $\bullet \otimes_R M$ to the sequence, what happens? Well, this time we get an exact sequence

$$A_1 \otimes_R M \longrightarrow A_2 \otimes_R M \longrightarrow A_3 \otimes_R M \longrightarrow 0.$$

Exercise 6.3.6. Prove this, and find an example showing that the leftmost map need not be an inclusion; $\bullet \otimes_R M$ is a *right exact functor*. ◇

Exercise 6.3.7. For R–modules M, N, and P, prove that

$$Hom_R(M \otimes_R N, P) \simeq Hom_R(M, Hom_R(N, P)).$$

Hint: Let $\phi \in Hom_R(M \otimes_R N, P)$. Given $m \in M$, we want to produce an element of $Hom_R(N, P)$. This is relatively easy: $\phi(m \otimes \bullet)$ expects elements of N as input and returns elements of P as output. Formalize this, and check that it is really an isomorphism. ◇

Given R-modules M_1 and M_2 and presentations as in the paragraph following Exercise 6.2.5, how do we obtain a presentation matrix for $M_1 \otimes_R M_2$? Clearly we have

$$R^{a_0} \otimes_R R^{b_0} \longrightarrow M_1 \otimes_R M_2 \longrightarrow 0,$$

so the question is how to compute the kernel. There are some obvious components, for example the submodules $im(\alpha) \otimes_R R^{b_0}$ and $R^{a_0} \otimes_R im(\beta)$. Can there be more?

Exercise 6.3.8. Given presentations for two R-modules M_1 and M_2, find a presentation for $M_1 \otimes_R M_2$. First look at the example below. ◇

Example 6.3.9.
```
i1 : R=ZZ/101[x,y,z];

i2 : M1 = coker matrix {{x,0,0},{0,y,0},{0,0,z}}

o2 = cokernel | x 0 0 |
              | 0 y 0 |
              | 0 0 z |
```

```
                          3
o2 : R-module, quotient of R

i3 : M2=coker transpose matrix {{x^3,y^3}}

o3 = cokernel {-3} | x3 |
              {-3} | y3 |

                          2
o3 : R-module, quotient of R

i4 : M1**M2

o4 = cokernel {-3} | x3 0  0   x 0 0 0 0 0 |
              {-3} | y3 0  0   0 0 0 x 0 0 |
              {-3} | 0  x3 0   0 y 0 0 0 0 |
              {-3} | 0  y3 0   0 0 0 0 y 0 |
              {-3} | 0  0  x3  0 0 z 0 0 0 |
              {-3} | 0  0  y3  0 0 0 0 0 z |

                          6
o4 : R-module, quotient of R
```

Supplemental Reading: Atiyah–Macdonald is good for all these topics,
I also like Munkres' treatment of tensor product. As always, Eisenbud has
everything.

Chapter 7

Geometry of Points and the Hilbert Function

Suppose that X consists of three distinct points in $\mathbb{P}^2$. For $i \gg 0$ we know that

$$HF(R/I(X), i) = HP(R/I(X), i) = 3.$$

The Hilbert polynomial tells us both the dimension and degree of X. However, there may be interesting geometric information "hiding" in the small values of the Hilbert function, that is, in those values where the Hilbert polynomial and Hilbert function don't agree. For example, we can't tell from the Hilbert polynomial if the three points are collinear or not. This is where the Hilbert function comes into its own. In particular, if the three points are collinear, then $HF(R/I(X), 1) = 2$, whereas if the three points are not collinear, then $HF(R/I(X), 1) = 3$. To see this, just note that if the points are collinear, then there is a linear form that vanishes on X, so the degree one piece of $I(X)$ has dimension one. On the other hand, if the points are not collinear, clearly the degree one piece of $I(X)$ is empty. In fact, this is just the tip of the iceberg; there are many beautiful connections between the Hilbert function and geometry, which we explore in this chapter.

Key concepts: Independent conditions, regularity, H^1, Macaulay–Gotzmann theorems, saturation, Hilbert difference function.

7.1 Hilbert Functions of Points, Regularity

To streamline notation, in the remainder of this chapter we write I_X for $I(X)$. For an arbitrary $X \subseteq \mathbb{P}^r_k$, $HF(I_X, i) = dim_k(I_X)_i$, which is the dimension of the vector space of degree i polynomials vanishing on X. If we are given a set of distinct points X, how can we compute the Hilbert function? Let's try another example in $\mathbb{P}^2$–in fact, let's find $HF(I_X, 2)$ when X consists of two points. A quadratic polynomial will have the form

$$f(x_0, x_1, x_2) = a_0 x_0^2 + a_1 x_0 x_1 + a_2 x_0 x_2 + a_3 x_1^2 + a_4 x_1 x_2 + a_5 x_2^2.$$

If $X = \{(b_0 : b_1 : b_2), (c_0 : c_1 : c_2)\}$, then $f(b_0 : b_1 : b_2) = 0$ if and only if

$$a_0 b_0^2 + a_1 b_0 b_1 + a_2 b_0 b_2 + a_3 b_1^2 + a_4 b_1 b_2 + a_5 b_2^2 = 0,$$

and similarly for $(c_0 : c_1 : c_2)$. So $f \in I_X$ if and only if $(a_0, a_1, a_2, a_3, a_4, a_5)$ is in the kernel of the matrix

$$\begin{bmatrix} b_0^2 & b_0 b_1 & b_0 b_2 & b_1^2 & b_1 b_2 & b_2^2 \\ c_0^2 & c_0 c_1 & c_0 c_2 & c_1^2 & c_1 c_2 & c_2^2 \end{bmatrix}.$$

In other words, finding $HF(I_X, i)$ for a given set of points $X = \{p_1, \ldots, p_n\}$ comes down to computing the rank of a matrix ϕ, where ϕ maps the vector space R_i to k^n by evaluation. If we choose a basis for R_i, the mth row of ϕ corresponds to evaluating the basis elements on p_m. The kernel of ϕ is $(I_X)_i$, so there is an exact sequence of vector spaces:

$$0 \longrightarrow (I_X)_i \longrightarrow R_i \overset{\phi}{\longrightarrow} k^n \longrightarrow H^1(I_X(i)) \longrightarrow 0,$$

where $H^1(I_X(i))$ denotes the cokernel of ϕ. We know that when $i \gg 0$ the Hilbert function of R/I is n, so this means that eventually (remember Exercise 2.3.1!) $H^1(I_X(i))$ must vanish. In math lingo, we say that a set X of n points imposes m conditions on polynomials of degree i if the rank of ϕ is m; if $n = m$ then we say that X imposes independent conditions on polynomials of degree i.

Exercise 7.1.1. Write a Macaulay 2 script (without using the command `hilbertFunction`!) which takes a set of points in $\mathbb{P}^2$ and a degree i, and computes the Hilbert function. Hint: determine the rank of the matrix obtained by evaluating the monomials of degree i at the points. To get the monomials of degree i, use the command `basis(i,R)`. To evaluate at a point $(b_0 : b_1 : b_2)$ you may find the command `map(R,R,{b_0,b_1,b_2})` useful. $\diamond$

Exercise 7.1.2. Do three collinear points in $\mathbb{P}^2$ impose independent conditions on linear forms? What if the points are not collinear? For 4, 5, and 6 points in $\mathbb{P}^2$, write down all possible Hilbert functions (up to $i =$ number of points). As an example, here are some of the cases that arise for 5 points:

Five collinear points:

degree i	=	0	1	2	3	4	5
$HF(R/I, i)$	=	1	2	3	4	5	5
$dim_k H^1(I_X(i))$	=	4	3	2	1	0	0

Four points on a line L, one point off L:

$$
\begin{array}{llllllll}
\text{degree } i & = & 0 & 1 & 2 & 3 & 4 & 5 \\
HF(R/I,i) & = & 1 & 3 & 4 & 5 & 5 & 5 \\
dim_k H^1(I_X(i)) & = & 4 & 2 & 1 & 0 & 0 & 0
\end{array}
$$

Five generic points:

$$
\begin{array}{llllllll}
\text{degree } i & = & 0 & 1 & 2 & 3 & 4 & 5 \\
HF(R/I,i) & = & 1 & 3 & 5 & 5 & 5 & 5 \\
dim_k H^1(I_X(i)) & = & 4 & 2 & 0 & 0 & 0 & 0
\end{array}
$$

When there are six points and the degree is 2, there are configurations X where no three points are collinear but the rank of ϕ is less than six. Describe this case. Find a configuration of points which has the same Hilbert function as X, but where some points are collinear. This exercise can be done by hand, but you will save time by using a computer. The scripts below (contained in a file called points) expect a set of points X as input; they return the ideal I_X. We demonstrate how to do the case of five collinear points.

```
pointideal1 = (m)->(v=transpose vars R;
            minors(2, (v|m)))
--return the ideal of the point
--represented by a column matrix

pointsideal1 = (m)->(
    t=rank source m;
    J=pointideal1(submatrix(m, ,{0}));
    scan(t-1, i->(J=intersect(J,
      pointideal1(submatrix(m, ,{i+1}))))));
    J)
--for a matrix with columns representing points,
--return the ideal of all the points.

i1 : load "points";

i2 :  R=ZZ/31991[x,y,z];

i3 : s1=random(R^2,R^5)

o3 = | 9534 14043 363   405    -10204 |
     | 7568 11665 5756 -6195  156    |
```

```
                 2       5
o3 : Matrix R   <--- R

i4 : s2=matrix{{0,0,0,0,0}}**R;

                 1       5
o4 : Matrix R   <--- R

i5 : s1||s2

o5 = | 9534 14043 363   405    -10204 |
     | 7568 11665 5756 -6195  156     |
     | 0    0     0     0      0      |

                 3       5
o5 : Matrix R   <--- R

i6 : I = pointsideal1 o5
```

$$o6 = \text{ideal } (z, \ x^5 + 2461x^4 y + 3083x^3 y^2 - 15951x^2 y^3 + 4853x*y^4 + 9253y^5)$$

```
o6 : Ideal of R

i7 : apply(10, i->hilbertFunction(i, coker gens I))

o7 = {1, 2, 3, 4, 5, 5, 5, 5, 5, 5}

i8 : betti res image gens I

o8 =   total: 2 1
           1: 1 .
           2: . .
           3: . .
           4: . .
           5: 1 1   ◇
```

It turns out that the vector spaces $H^1(I_X(i))$ are related in a fundamental way to the free resolution of I_X, so it is important to determine when they vanish. Go back and compute the free resolutions for the ideals you found in the previous exercise and see if you can come up with a conjecture relating the number of rows in the Macaulay 2 betti diagram to the vanishing of $H^1(I_X(i))$.

Lemma 7.1.3. *A set of points* $X = \{p_1, \ldots, p_n\} \subseteq \mathbb{P}^r$ *imposes* m *conditions on polynomials of degree* i *iff there exists* $Y \subseteq X, |Y| = m$ *such that for each* $p \in Y$ *there exists a polynomial of degree* i *which vanishes on* $Y - p$ *but is nonzero on* p. *Such polynomials are said to separate the points of* Y.

Proof. X imposes m conditions when the rank of ϕ is at least m. But we can just do a column reduction (which corresponds to choosing a new basis for the space of degree i polynomials) and then permute rows so that ϕ has upper left block an m by m identity matrix, done. $\square$

Lemma 7.1.4. *If* $X = \{p_1, \ldots, p_n\} \subseteq \mathbb{P}^r$ *imposes* $m < n$ *conditions on polynomials of degree* i, *then* X *imposes at least* $m + 1$ *conditions on polynomials of degree* $i + 1$.

Proof. Let $Y \subseteq X$ be as in the previous lemma. Pick $p \in X - Y$, and set $Y' = Y + p$. Pick a linear form l such that $l(p) = 0, l(q) \neq 0$ for all $q \in Y$ (we can do this if our field is reasonably big relative to the number of points–intuitively, for a finite set of points, you can always find a hyperplane through one and missing the rest–draw a picture!). If $\{f_1, \ldots, f_m\}$ are the polynomials of degree i which separate the points of Y, then $\{l \cdot f_1, \ldots, l \cdot f_m\}$ still separate the points of Y, and are all zero on p. So we need a polynomial of degree $i + 1$ which is zero on Y and not at p. Since $p \notin Y$ we can find a polynomial which is zero on Y and nonzero on p. But the ideal of Y is generated in degree at most i, so we can actually find such a thing of degree $i + 1$. $\square$

Exercise 7.1.5. Prove that the ideal of Y is generated in degree at most i. ◇

Lemma 7.1.6. *If* $H^1(I_X(i-1))$ *is nonzero, then* $dim_k H^1(I_X(i-1)) > dim_k H^1(I_X(i))$. $H^1(I_X(i)) = 0$ *if* $i \geq |X| - 1$.

Proof. The first part follows immediately from the previous lemma. For the second part, just observe that the biggest possible value for $H^1(I_X(1))$ is $|X| - 2$, which occurs only if the points of X are all collinear. $\square$

Definition 7.1.7. *Let M be a finitely generated, graded module over* $R = k[x_0, \ldots, x_r]$ *with minimal free resolution*

$$0 \longrightarrow F_{r+1} \longrightarrow \cdots \longrightarrow F_0 \longrightarrow M \longrightarrow 0,$$

where

$$F_i \simeq \bigoplus_j R(-a_{i,j}).$$

The regularity of M is

$$\sup_{i,j}\{a_{i,j} - i\}.$$

For example, the regularity of the ideal I_X of five points on a line in $\mathbb{P}^2$ is five since the free resolution of I_X is given by:

$$0 \longrightarrow R(-6) \longrightarrow R(-5) \oplus R(-1) \longrightarrow I_X \longrightarrow 0.$$

Flip back and look at the output of the Macaulay 2 command betti for this example: the regularity of I_X is displayed as the number at the left of the bottommost row. We defer the proof of the following theorem till Chapter 10:

Theorem 7.1.8. *For a set of points X in* $\mathbb{P}^n$*, the regularity of* I_X *is the smallest i such that* $H^1(I_X(i-1))$ *vanishes.*

Exercise 7.1.9. Prove that for a generic set of n points X in $\mathbb{P}^r$, the betti diagram has at most two nonzero rows, which are adjacent. Phrased a bit differently, if i is the largest integer such that

$$n > \binom{r+i-1}{i-1},$$

then I_X is $i + 1$ regular. Lets check the example of 5 points in $\mathbb{P}^2$:

$$5 > \binom{2+2-1}{2-1}, \ 5 \not> \binom{2+3-1}{3-1},$$

so the ideal of five generic points in $\mathbb{P}^2$ is $2 + 1 = 3$ regular. How about $20 = \binom{3+3}{3}$ generic points in $\mathbb{P}^3$? The points impose 20 conditions on cubics, so (since there are no cubics in the ideal) I_X is first nonzero in degree four. On the other hand, the map ϕ is full rank (20) when $i = 3$, so four is also the smallest i such that $H^1(I_X(i-1)) = 0$. Thus, for this example, we expect the resolution to occur in a *single* row. Here is the resolution for I:

```
i1 : R=ZZ/31991[x,y,z,w]; load "points"

i2 : M = random(R^4,R^20);

                4       20
o2 : Matrix R   <--- R

i3 : I = pointsideal1 M;

o3 : Ideal of R

i4 : betti res coker gens I

o4 = total: 1 15 24 10
          0: 1  .   .   .
          1: .   .   .   .
          2: .   .   .   .
          3: .  15 24 10

i5 : betti res image gens I

o5 = total: 15 24 10
          4: 15 24 10
```

Hint: from the definition of generic, how many conditions does X impose in each degree, and what does this mean about the rank of ϕ? Also, notice that the regularity of I is one more than the regularity of R/I. Using Definition 7.1.7, prove this is always the case. ◇

Exercise 7.1.10. Write a program which takes two parameters n and r, and generates the betti numbers of a set of i generic points in $\mathbb{P}^r$, for all $r \le i \le n$. You will need to use a loop structure (try either apply or scan), and the random command. By the previous exercise there will only be two nonzero rows in the betti diagram for I. Run your program for $n = 12$ and all $r \le 5$, and make a conjecture about the shape of the free resolutions. You should come up with the *minimal resolution conjecture* – see Lorenzini [62]. Now do this for $r = 6$. What happens? In [36] Eisenbud and Popescu use *Gale duality* to understand this example (due to Schreyer) and in fact provide infinite families with similar behavior. ◇

7.2 The Theorems of Macaulay and Gotzmann

We now record two fundamental theorems on Hilbert functions:

Theorem 7.2.1 (Macaulay). *Let $\{h_0, h_1, \ldots\}$ be a sequence of non-negative integers, and write the i-binomial expansion of h_i as*

$$h_i = \binom{a_i}{i} + \binom{a_{i-1}}{i-1} + \cdots$$

There is a unique way to do this if we require that $a_i > a_{i-1} > \cdots$. Define

$$h_i^{\langle i \rangle} := \binom{a_i + 1}{i+1} + \binom{a_{i-1} + 1}{i} + \cdots .$$

Then $\{h_0, h_1, \ldots\}$ is the Hilbert function of a $\mathbb{Z}$-graded algebra over a field iff for all $i \geq 1$, $h_{i+1} \leq h_i^{\langle i \rangle}$ and $h_0 = 1$.

Macaulay says in his paper that the proof is "too long and complicated to provide any but the most tedious reading"; a shorter and simpler proof may be found in [46].

Example 7.2.2. The 5-binomial expansion of 75 is

$$75 = \binom{8}{5} + \binom{6}{4} + \binom{4}{3}.$$

Suppose $h_5 = 75$, then

$$h_5^{\langle 5 \rangle} = \binom{9}{6} + \binom{7}{5} + \binom{5}{4} = 110.$$

So if the Hilbert function of a $\mathbb{Z}$-graded k-algebra A satisfies $HF(A, 5) = 75$, then $HF(A, 6) \leq 110$.

A theorem of Gotzmann says what happens when we have an equality in the above theorem:

Theorem 7.2.3 (Gotzmann). *If I is generated in degree i, and if $h_{i+1} = h_i^{\langle i \rangle}$, then for any $j \geq 1$*

$$h_{i+j} = \binom{a_i + j}{i+j} + \binom{a_{i-1} + j - 1}{i+j-1} + \cdots .$$

The condition that $h_{i+1} = h_i^{\langle i \rangle}$ means that I grows as little as possible in going from degree i to degree $i + 1$. Roughly speaking, Gotzmann's theorem says that if an ideal generated in degree i has stunted growth in degree $i + 1$, then it is forever stunted. Notice that Gotzmann's theorem also gives us the Hilbert polynomial of R/I.

Exercise 7.2.4. Which of the following are Hilbert functions of $\mathbb{Z}$-graded k-algebras?

1. $(1,4,6,8,19,4,5,1,0)$
2. $(1,16,15,14,13,12,6,0)$
3. $(1,3,6,4,1,2,1)$ $\diamond$

7.3 Artinian Reduction and Hypersurfaces

Suppose a set of points X contains a large subset Y lying on a hypersurface of low degree. If L is a generic linear form, then the Hilbert function of the *Artinian reduction* $R/\langle I_X, L \rangle$ can often "see" Y. We begin with an example:

Example 7.3.1. Let X consist of 24 points in $\mathbb{P}^2$. Suppose twenty of the points lie on the conic C:

$$xz - y^2 = 0,$$

and the remaining four points are in generic position. On the open patch U_x we can set $x = 1$, hence a point on the conic will have coordinates $(1 : y : y^2)$. We can use our code from section one to obtain I_X:

```
i1 : R=ZZ/31991[x,y,z]; load "points"

o2 : VDM = (a,b)-> (map(ZZ^a,ZZ^b, (i,j)-> (j+1)^i))

   --return an a by b Van Der Monde matrix

i3 : A=VDM(3,20)**R

o3 = {0} | 1 1 1 1  1  1  1  ... 1  |
     {0} | 1 2 3 4  5  6  7  ... 20 |
     {0} | 1 4 9 16 25 36 49 ... 400 |

                3         20
o3 : Matrix R  <--- R

i4 : B= random(R^3,R^4)

o4 = {0} | 9534  11665 405    156    |
     {0} | 7568  363   -6195  -14844 |
     {0} | 14043 5756  -10204 -2265  |
```

```
                 3         4
o4 : Matrix R   <--- R

i5 : betti res coker gens pointsideal1 (A|B)

o5 = total: 1 3 2
           0: 1 . .
           1: . . .
           2: . . .
           3: . 2 .
           4: . . 1
           5: . . .
           6: . . .
           7: . . .
           8: . . .
           9: . 1 .
          10: . . 1

o5 : Net

i6 : apply(12, i->hilbertFunction(i,
     coker gens pointsideal1 (A|B)))

o6 = {1, 3, 6, 10, 13, 15, 17, 19, 21, 23, 24, 24}

i7 : K=(pointsideal1(A|B))+ideal random(R^1,R^{-1});
     --add a random linear form

i8 : apply(12, i->hilbertFunction(i, coker gens K))

o8 = {1, 2, 3, 4, 3, 2, 2, 2, 2, 2, 1, 0}
```

The long string of twos in the Hilbert function of $R/\langle I_X, L \rangle$ looks interesting. What is happening is the following: the four generic points of X impose four conditions on conics, so there are two conics Q_1, Q_2 passing through those four points. Hence, the quartics CQ_1, CQ_2 pass through all the points, and as we see from the betti diagram $(I_X)_4$ contains two elements, which are the quartics we found. The long string of two's in the Hilbert function of $R/\langle I_X, L \rangle$ is closely related to the fact that the Hilbert function of R/I_X has maximal growth in passing from degree eight to degree nine. These numerical phenomena reflect the fact that the elements of $(I_X)_4$ have a greatest common divisor (the conic C).

So a natural question is: what conditions force I_d to have a greatest common divisor? Consider three quadratic polynomials in three variables: if they are general, then the syzygies are Koszul, so that $HF(R/I, 3) = 1$. On the other hand, if the quadratics share a common factor, say $I = \langle x^2, xy, xz \rangle$, then there are 3 linear syzygies, so $HF(R/I, 3) = 4$. Check that this is the maximal possible growth of R/I in passing from degree two to degree three. This suggests that the existence of a GCD for I_d might be linked to the maximal growth of the Hilbert function in degree d. In what follows we describe results of Davis [25], and generalizations due to Bigatti, Geramita and Migliore [15], which also has an excellent exposition and description of the history of this problem. We begin with the observation that if $g \in k[x_0, \dots, x_r]_j$, then

$$HF(k[x_0, \dots, x_r]/\langle g \rangle, d) = \binom{d+r}{r} - \binom{d-j+r}{r} =: f_{r,j}(d).$$

Exercise 7.3.2. Prove that $f_{r,j}(d) = \binom{d+r-1}{r-1} + \binom{d+r-2}{r-1} + \cdots + \binom{d+r-j}{r-1}$. ◇

Definition 7.3.3. *Let* $I \subset k[x_0, \dots, x_r] = R$ *be a homogeneous ideal with* $I_d \neq 0$. *The potential GCD degree of* I_d *is*

$$sup\{j \,|\, f_{r,j}(d) \leq HF(R/I, d)\}.$$

Theorem 7.3.4. *Let* $I \subset k[x_0, \dots, x_r]$ *be a homogeneous ideal with* $I_d \neq 0$, *and suppose* $0 < j$ *is the potential GCD degree of* I_d. *If* R/I *has maximal growth in degree* d, *then both* I_d *and* I_{d+1} *have a GCD of degree* j.

Proof. From the exercise above, the first j terms of $f_{r,j}(d)$ and $f_{r,j+1}(d)$ (expanded as binomial coefficients with $r - 1$ "denominator") are equal. Since

$$f_{r,j}(d) \leq HF(R/I, d) < f_{r,j+1}(d),$$

we can write

$$HF(R/I, d) = \binom{d+r-1}{r-1} + \binom{d+r-2}{r-1} + \cdots + \binom{d+r-j}{r-1}$$
$$+ \text{ terms } \binom{c}{i},$$

where $i < r - 1$ and $c < d + r - (j + 1)$. But now from the maximal growth assumption, Gotzmann's theorem implies that the Hilbert polynomial of $R/I_{\leq d}$ has degree $r - 1$ and lead coefficient $\frac{j}{(r-1)!}$, which is the Hilbert polynomial of a degree j hypersurface. This also forces a GCD in degree $d + 1$ because the minimal growth of I in passing to degree $d + 1$ means that I_{d+1}

is generated by I_d. So there can be no new minimal generators of I in degree $d + 1$ and a GCD for I_d forces a GCD for I_{d+1}. $\square$

It is important to note that if I_d has a GCD g, this need not imply that $V(I) \subseteq V(g)$ – this is illustrated by the example at the beginning of the section. However, what it does do is give us a natural way to "split" $V(I)$ into two parts: the part which is contained in the hypersurface $V(g)$, and the part off the hypersurface. As we just proved, the maximal growth of the Hilbert function is related to the existence of a GCD, but our example also showed that the GCD manifested itself in the Hilbert function of a hyperplane section. We now explore this more carefully. Recall that if $I \subseteq k[x_0, \ldots, x_r]$ is a homogeneous ideal, the dimension and degree of the projective variety $V(I)$ may be defined in terms of the Hilbert polynomial of R/I. For example, a curve in $\mathbb{P}^r$ is an object $V(I)$ such that

$$HP(R/I, i) = ai + b,$$

for some constants a and b. In Chapter 3, we saw that if the maximal ideal is not an associated prime of I, then there is a linear form L which is a nonzerodivisor on R/I, so we obtain an exact sequence

$$0 \longrightarrow R(-1)/I \xrightarrow{\cdot L} R/I \longrightarrow R/\langle I, L \rangle \longrightarrow 0.$$

Thus,

$$HP(R/\langle I, L \rangle, i) = ai + b - (a(i-1) + b) = a,$$

which means that $V(L)$ meets the curve in a points. We underscore the idea: it is often possible to understand an object by slicing it with hyperplanes, and then examining what those slices look like (remember level curves from calculus!). On the other hand, if $V(I)$ is a zero dimensional object in $\mathbb{P}^r$, then it might seem that this technique is useless: there is nothing left when we slice with a generic hyperplane, since

$$V(I, L) = \emptyset.$$

Although there is indeed no geometric object left, there is a little Artinian "residual" algebra

$$A = k[x_0, \ldots, x_r]/\langle I, L \rangle.$$

It turns out that lots of wonderful information is hiding in A! First, we need some background and notation. For a graded R-module M, define

$$\Delta H(M, i) = HF(M, i) - HF(M, i - 1),$$

so that if L is a nonzerodivisor on M, $\Delta H(M, i) = HF(M/LM, i)$. Next, we say a few words about a particularly important case of ideal quotient:

Example 7.3.5. Saturation

Consider the ideals

$$I = \langle x \rangle \text{ and } J = \langle x^3, x^2y, x^2z, xy^2, xyz, xz^2 \rangle.$$

A quick check shows that the primary decomposition of J is

$$\langle x \rangle \cap \langle x, y, z \rangle^3.$$

As we remarked earlier, the variety defined by any ideal primary to the maximal ideal is empty (in projective space), so if we want to study projective varieties by the method above, we have to be a bit careful. In particular, although having a component primary to the maximal ideal does not affect the Hilbert polynomial, it can considerably alter the Hilbert function.

$$
\begin{array}{rcccccccc}
\text{degree } i & = & 0 & 1 & 2 & 3 & 4 & 5 & 6 \\
HF(R/I, i) & = & 1 & 2 & 3 & 4 & 5 & 6 & 7 \\
HF(R/J, i) & = & 1 & 3 & 6 & 4 & 5 & 6 & 7
\end{array}
$$

So if we want to look at Hilbert functions, we need start by removing this ambiguity.

Definition 7.3.6. *Let I and J be ideals in a polynomial ring. Then $I : J^\infty = \{f \mid f \in I : J^n \text{ for some } n\}$.*

Since R is a Noetherian ring, the chain

$$I : J \subseteq I : J^2 \subseteq \cdots$$

must stabilize; the stable value $I : J^\infty$ is called the saturation of I with respect to J. If J is the ideal generated by the variables, then $I^{sat} = I : J^\infty$ is the *saturation* of I. It is the largest ideal such that $I^{sat}_i = I_i$ for $i \gg 0$.

Exercise 7.3.7. Let $I \subseteq k[x_0, \ldots, x_r]$ be a saturated ideal of codimension r.

1. Prove that there exists a linear form L such that

$$\sum_{i=0}^\infty HF(R/\langle I, L \rangle, i) = \sum_{i=0}^\infty \Delta H(R/I, i) = HP(R/I, t).$$

For example, if I is radical, then $\sum_{i=0}^{\infty} \Delta H(R/I, i)$ is the number of points of $V(I)$.

2. Show that for g of degree j, if $\langle I, g \rangle$ is also saturated, then the exact sequence

$$0 \longrightarrow R(-j)/\langle I : g \rangle \longrightarrow R/I \longrightarrow R/\langle I, g \rangle \longrightarrow 0,$$

gives rise to

$$\Delta H(R/I, i) = \Delta H(R/\langle I : g \rangle, i - j) + \Delta H(R/\langle I, g \rangle, i). \quad \diamond$$

When I_d has a GCD of degree j and $\Delta H(R/I, d) = f_{r,j}(d)$, the following theorem of Bigatti, Geramita and Migliore gives very specific information about how the Hilbert difference functions split up. As you might guess, the proof follows along the lines of the exercise above. See [15] for the details.

Theorem 7.3.8. *Let $I \subseteq k[x_0, \ldots, x_{r+1}]$ be a saturated ideal, and suppose that I_d has a greatest common divisor f of degree j, and that $\Delta H(R/I, d) = f_{r,j}(d)$. Then if we define*

$$I_{X_1} = \langle I, f \rangle \text{ and } I_{X_2} = \langle I : f \rangle,$$

we have that

$$
\begin{array}{rcll}
\Delta H(R/I_{X_1}, i) & = & f_{r,j}(t) & \text{if } i \leq d \\
\Delta H(R/I_{X_1}, i) & = & \Delta H(R/I_X, i) & \text{if } i > d \\
\Delta H(R/I_{X_2}, i - j) & = & \Delta H(R/I_X, i) - f_{r,j}(i) & \text{if } i \leq d \\
\Delta H(R/I_{X_2}, i - j) & = & 0 & \text{if } i > d
\end{array}
$$

Example 7.3.9. For the 24 points in Example 7.3.1 we have

$$f_{2,j}(d) = dj + \frac{3j - j^2}{2}.$$

Solving for j in

$$f_{2,j}(8) = 8j + \frac{3j - j^2}{2} \leq HF(R/I, 8) = 21,$$

we see that $j = 2$ is the potential GCD degree. We compute

$$21 = \binom{9}{8} + \binom{8}{7} + \binom{6}{6} + \binom{5}{5} + \binom{4}{4} + \binom{3}{3}$$

so $HF(R/I, 9) \leq 23$, hence R/I has maximal growth in degree eight. By Theorem 7.3.4 I_8 has a GCD C of degree two. Now, $f_{1,2}(8) = 2 = \Delta H(R/I, 8)$, so Theorem 7.3.8 tells us that for the subset X_1 of points on C

degree i	=	0	1	2	3	4	5	6	7	8	9	10
$\Delta H(R/I_{X_1}, i)$	=	1	2	2	2	2	2	2	2	2	2	1

So if we had started out with no knowledge about X except the Hilbert function, we would have been able to determine that 20 of the points of X lie on a conic! Use your code to come up with some interesting examples and experiment. The paper [15] also contains interesting results on sets of points with the *uniform position property* (see [33]) and on sets of points with "many" points on a subvariety (see also [73]).

Exercise 7.3.10. If $\Delta H(R/I, d) = f_{r,j}(d)$ and $\Delta H(R/I, d+1) = f_{r,j}(d+1)$, prove that I_d and I_{d+1} have a greatest common divisor of degree j. In particular, Theorem 7.3.8 applies. $\diamond$

Supplemental Reading: For more on regularity, see Chapter 4 of [29] or Chapter 18 of [28]. The study of points in $\mathbb{P}^n$ is a very active field with a vast literature. An excellent introduction to the area may be found in the Queen's Papers articles by Tony Geramita [43], [44]; other nice expositions are by Harbourne [51] and Miranda [70]. I learned the approach of Section 7.1 from Sorin Popescu. In [31] Eisenbud, Green and Harris give a wonderful introduction to the "Cayley-Bacharach problem", which concerns collections of points (more properly, subschemes) lying on a complete intersection.

If we allow points to occur with multiplicity, then they are called "fat points", and there are surprising connections to ideals generated by powers of linear forms and secant varieties; for this see the work of Emsalem and Iarrobino [37] and Iarrobino and Kanev [57]. A final topic worth mentioning is the notion of the Hilbert scheme: the basic idea is to study all varieties of $\mathbb{P}^n$ having the same Hilbert polynomial at once. A priori this is just some set, but in fact there is *lots* of algebraic structure, and Hilbert schemes for zero-dimensional objects have received quite a bit of attention. A very quick introduction to the Hilbert scheme can be found in the book of Smith, Kahanpää, Kekäläinen, Traves [86] (which is another nice elementary algebraic geometry text); or in the Montreal lecture notes of Eisenbud and Harris [33].

Chapter 8

Snake Lemma, Derived Functors, Tor and Ext

In Chapter 2, we introduced the concept of a *chain complex*, which is a sequence of modules and homomorphisms such that the image of one map is contained in the kernel of the next map. Free resolutions are one very specific type of chain complex: the modules are free and the sequence is actually exact; in Chapter 5 we saw that there are also many interesting chain complexes which are not of this type. We now introduce two key tools used to study chain complexes – the snake lemma, and the long exact sequence in homology. We also take a look at the topological roots and motivation, specifically, the Mayer–Vietoris sequence. We saw in Chapter 6 that for an arbitrary R–module M the functor $\bullet \otimes_R M$ is right exact but not (in general) exact. In this situation, we define objects called *higher derived functors* which measure the failure of exactness. This relates to free resolutions and chain complexes because the higher derived functors of $\bullet \otimes_R M$ can be computed via free resolutions; the long exact sequence in homology plays a key role. One beautiful payoff is that this abstract machinery actually gives a *really slick* proof of the Hilbert syzygy theorem!

Key concepts: Snake lemma, long exact sequence, Mayer–Vietoris sequence, derived functors, Tor, Ext.

8.1 Snake Lemma, Long Exact Sequence in Homology

Suppose we have a commutative diagram of R-modules with exact rows:

$$
\begin{array}{ccccccccc}
0 & \longrightarrow & A_1 & \xrightarrow{a_1} & A_2 & \xrightarrow{a_2} & A_3 & \longrightarrow & 0 \\
 & & \downarrow f_1 & & \downarrow f_2 & & \downarrow f_3 & & \\
0 & \longrightarrow & B_1 & \xrightarrow{b_1} & B_2 & \xrightarrow{b_2} & B_3 & \longrightarrow & 0
\end{array}
$$

Are the kernels and cokernels of the vertical maps related? Indeed they are!

Lemma 8.1.1 (The Snake Lemma). *For a diagram as above, we have an exact sequence:*

$$0 \longrightarrow kernel\ f_1 \longrightarrow kernel\ f_2 \longrightarrow kernel\ f_3 \overset{\phi}{\longrightarrow}$$
$$cokernel\ f_1 \longrightarrow cokernel\ f_2 \longrightarrow cokernel\ f_3 \longrightarrow 0.$$

Proof. The key is defining the map ϕ from kernel f_3 to cokernel f_1. Let α_3 be in the kernel of f_3. Since a_2 is surjective, we can find $\alpha_2 \in A_2$ such that $a_2(\alpha_2) = \alpha_3$. Then $f_2(\alpha_2) \in B_2$. Furthermore, since the diagram commutes, we have that

$$b_2(f_2(\alpha_2)) = f_3(a_2(\alpha_2)) = 0,$$

so $f_2(\alpha_2)$ is in the kernel of b_2. By exactness there is a $\beta_1 \in B_1$ such that $b_1(\beta_1) = f_2(\alpha_2)$. We put $\phi(\alpha_3) = \beta_1$. Let's check that this is well defined. Suppose we chose α_2' mapping to α_3. Then $\alpha_2 - \alpha_2'$ is in the kernel of a_2, so there exists $\alpha_1 \in A_1$ with

$$a_1(\alpha_1) = \alpha_2 - \alpha_2',$$

so $\alpha_2 = \alpha_2' + a_1(\alpha_1)$. Now we map things forward, so

$$f_2(\alpha_2) = f_2(\alpha_2' + a_1(\alpha_1)) = f_2(\alpha_2') + b_1 f_1(\alpha_1).$$

So β_1 is well defined *as an element of coker* f_1, because $f_2(\alpha_2) = f_2(\alpha_2')$ mod the image of f_1. $\square$

Exercise 8.1.2. Finish the proof of the Snake lemma by showing exactness. With the Snake in hand, give a short proof of Exercise 2.3.5.2. ◇

Exercise 8.1.3. For a short exact sequence of modules

$$0 \longrightarrow A_1 \overset{a_1}{\longrightarrow} A_2 \overset{a_2}{\longrightarrow} A_3 \longrightarrow 0,$$

show that $A_2 \simeq A_1 \oplus A_3$ iff there is a homomorphism b_2 with $a_2 b_2 = id_{A_3}$ or a homomorphism b_1 with $b_1 a_1 = id_{A_1}$. ◇

A *short exact sequence of complexes* is a diagram:

$$
\begin{array}{ccccccc}
 & & 0 & & 0 & & 0 \\
 & & \downarrow & & \downarrow & & \downarrow \\
A: \cdots & \longrightarrow & A_2 & \overset{\partial_2}{\longrightarrow} & A_1 & \overset{\partial_1}{\longrightarrow} & A_0 \longrightarrow 0 \\
 & & \downarrow & & \downarrow & & \downarrow \\
B: \cdots & \longrightarrow & B_2 & \overset{\partial_2}{\longrightarrow} & B_1 & \overset{\partial_1}{\longrightarrow} & B_0 \longrightarrow 0 \\
 & & \downarrow & & \downarrow & & \downarrow \\
C: \cdots & \longrightarrow & C_2 & \overset{\partial_2}{\longrightarrow} & C_1 & \overset{\partial_1}{\longrightarrow} & C_0 \longrightarrow 0 \\
 & & \downarrow & & \downarrow & & \downarrow \\
 & & 0 & & 0 & & 0
\end{array}
$$

where the columns are exact and the rows are complexes.

Theorem 8.1.4 (Long Exact Sequence in Homology). *A short exact sequence of complexes yields a long exact sequence in homology, i.e. an exact sequence*

$$\cdots \longrightarrow H_{n+1}(C) \longrightarrow H_n(A) \longrightarrow H_n(B) \longrightarrow H_n(C) \longrightarrow H_{n-1}(A) \longrightarrow \cdots$$

Exercise 8.1.5. Prove the previous theorem. The Snake lemma gives the theorem for an exact sequence of complexes where each complex is of the form

$$0 \longrightarrow F_1 \longrightarrow F_0 \longrightarrow 0.$$

Now induct. ◇

Homological algebra has its roots in topology, and we illustrate the usefulness of the above theorem with a topological application. Suppose we glue two simplicial complexes X and Y together along a common subcomplex $Z = X \cap Y$. What is the relationship between the homology of $X \cup Y$, $X \cap Y$ and X and Y? Let C_i denote the oriented i-simplices. We have a surjective map:

$$C_i(X) \oplus C_i(Y) \xrightarrow{g} C_i(X \cup Y),$$

defined by $g(\sigma_1, \sigma_2) = \sigma_1 + \sigma_2$. Notice that if we take a simplex $\sigma \in X \cap Y$, then $g(\sigma, -\sigma) = 0$, so that the kernel of g is isomorphic to $C_i(X \cap Y)$. Thus, we get a short exact sequence of complexes

$$0 \longrightarrow C_\bullet(X \cap Y) \longrightarrow C_\bullet(X) \oplus C_\bullet(Y) \longrightarrow C_\bullet(X \cup Y) \longrightarrow 0.$$

From the long exact sequence in homology, we obtain the *Mayer–Vietoris* sequence:

$$\cdots \longrightarrow H_{n+1}(X \cup Y) \longrightarrow H_n(X \cap Y) \longrightarrow H_n(X) \oplus H_n(Y)$$
$$\longrightarrow H_n(X \cup Y) \longrightarrow H_{n-1}(X \cap Y) \longrightarrow \cdots$$

Example 8.1.6. Suppose we want to compute the homology with $\mathbb{Q}$-coefficients of the torus with a puncture. In other words, we're going to remove a little triangular patch. Let X play the role of the punctured torus and Y be the little triangular patch. So we have that $X \cup Y$ is the torus T^2, and $X \cap Y$ is a little (hollow) triangle $\simeq S^1$. Mayer–Vietoris gives us the exact sequence:

$$0 \longrightarrow H_2(S^1) \longrightarrow H_2(X) \oplus H_2(Y) \longrightarrow H_2(T^2) \longrightarrow$$
$$H_1(S^1) \longrightarrow H_1(X) \oplus H_1(Y) \longrightarrow H_1(T^2) \longrightarrow$$
$$H_0(S^1) \longrightarrow H_0(X) \oplus H_0(Y) \longrightarrow H_0(T^2) \longrightarrow 0$$

We know the homology of everything in sight except X:

	$dim(H_0)$	$dim(H_1)$	$dim(H_2)$
S^1	1	1	0
T^2	1	2	1
Y	1	0	0

Since X is connected, $dim(H_0(X)) = 1$, which means (by exactness) that the map from $H_1(T^2)$ to $H_0(S^1)$ is the zero map. So since $dim(H_1(Y)) = 0$ this means that $dim(H_1(X)) - dim(H_2(X)) = 2$. That is all we can get from the exact sequence. Remove a two-simplex from the triangulation that appears in Exercise 5.1.8. This yields a triangulation of X. Verify that $H_2(X) = 0$, hence $dim(H_1(X)) = 2$.

Exercise 8.1.7. Identify all the vertices in the picture below, and identify the edges according to the labelling. Call the resulting object X_2.

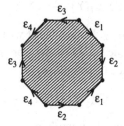

The underlying topological space of X_2 is a "sphere with two handles" or more formally a real, compact, connected, orientable surface of genus 2.

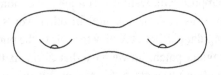

Take two copies of the simplicial complex corresponding to the punctured torus of Example 8.1.6 and identify them along the boundary of the "missing" triangle (make sure the orientations of the boundaries of the "missing triangles" agree). Topologically this corresponds to gluing together pieces as below.

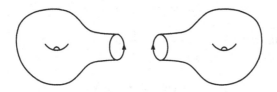

Compute the homology of X_2. You should obtain $dim(H_0) = dim(H_2) = 1$, and $dim(H_1) = 4$. Prove that for a simplicial complex X_g whose underlying topological space is a real, compact, connected, orientable surface of genus g that $dim(H_0) = dim(H_2) = 1$, and $dim(H_1) = 2g$. Recall that Poincaré duality says $dim(H_i) = dim(H_{n-i})$ for a certain class of "nice" n-dimensional orientable manifolds; we're seeing that here. Naively, orientable means that for a surface S sitting in $\mathbb{R}^3$ you can choose a normal vector at a point $p \in S$, then travel around S (dragging the normal vector with you); when you get back to p, the normal points the same way as at the start of the trip. The Möbius band is not orientable, nor is the Klein bottle of Exercise 5.1.8. For more on the classification of surfaces and the formal definition of orientability, see [41]. $\diamond$

8.2 Derived Functors, Tor

Let R be (as always) a commutative ring with unit and C be the category of R-modules and R-module homomorphisms. We know that some functors from C to C do not preserve exactness: if we apply $\bullet \otimes_R M$ to a short exact sequence

$$0 \longrightarrow A_1 \longrightarrow A_2 \longrightarrow A_3 \longrightarrow 0,$$

then $A_1 \otimes_R M \longrightarrow A_2 \otimes_R M$ may have a kernel. We're going to define a collection of objects which measure, in some sense, the failure of a functor to be exact. Although the initial definition seems a little forbidding, we'll see that

- *you can compute these things*
- *they yield lots of useful information*

Definition 8.2.1. *Let F be a right exact, covariant, additive (preserves addition of homomorphisms) functor from C to C. Then for an object N we define* *the left derived functors $L_i F(N)$ as follows: Take a projective resolution $P_\bullet$ for N:*

$$\cdots \longrightarrow P_2 \xrightarrow{d_2} P_1 \xrightarrow{d_1} P_0 \longrightarrow N \longrightarrow 0.$$

Apply F to the complex $\cdots \longrightarrow P_2 \xrightarrow{d_2} P_1 \xrightarrow{d_1} P_0 \longrightarrow 0$, *and call the resulting complex $F_\bullet(N)$. Then*

$$L_i F(N) = H_i(F_\bullet(N)).$$

In words, the ith homology of $F_\bullet(N)$ is the ith left derived functor of F. $F_\bullet(N)$ is a complex because F preserves composition, so $d_i d_{i+1} = 0$ means $0 = F(d_i d_{i+1}) = F(d_i)F(d_{i+1})$. Our paradigm for F is $\bullet \otimes_R M$.

Exercise 8.2.2. (Chain Homotopy). Let $(A_\bullet, d)$ and $(B_\bullet, \delta)$ be chain complexes, and let α, β be *chain maps*: for each i, α_i and β_i are homomorphisms from A_i to B_i which commute with the differentials. If there exists a family γ of homomorphisms $A_i \xrightarrow{\gamma_i} B_{i+1}$ with $\alpha - \beta = \delta\gamma + \gamma d$ then α and β are *homotopic*.

1. Prove that a chain map induces a map on homology (easy).
2. Prove that if α and β are homotopic, then they induce the same map on homology (easy).
3. For F as in Definition 8.2.1, show that $L_i F(N)$ does not depend on the projective resolution chosen for N. Hint: take projective resolutions $P_\bullet$ and $Q_\bullet$ for N, and use projectivity to obtain chain maps between them which compose to give the identity. Then you'll need to use additivity. (harder!) If you get stuck, see [28], Corollary A.3.14. ◇

Example 8.2.3. We work in the category of graded $k[x]$-modules and graded homomorphisms, and consider the functor

$$F = \bullet \otimes_{k[x]} k[x]/x.$$

Suppose we want to compute $L_i F(k[x]/x^2)$, for all i. Well, first we take a projective resolution for $k[x]/x^2$:

$$0 \longrightarrow k[x](-2) \xrightarrow{\cdot x^2} k[x] \longrightarrow k[x]/x^2 \longrightarrow 0.$$

Dropping $k[x]/x^2$ and applying F we obtain the complex

$$0 \longrightarrow k[x](-2) \otimes k[x]/x \xrightarrow{\cdot x^2} k[x] \otimes k[x]/x \longrightarrow 0.$$

Since $A \otimes_A M \simeq M$, this is

$$0 \longrightarrow k[x](-2)/x \xrightarrow{\cdot x^2} k[x]/x \longrightarrow 0.$$

Since $\cdot x^2$ is the zero map, the homology at position zero is just $k[x]/x$, i.e.

$$L_0 F(k[x]/x^2) = k[x]/x.$$

In fact, it is easy to see from the definition that the zeroth left derived functor of an object N is just $F(N)$. What about L_1? We have that $L_1 F(k[x]/x^2)$ is the kernel of the map $\cdot x^2$. But when we tensored with $k[x]/x$, we set x to zero, so this is the zero map: *everything* is in the kernel. So

$$L_1 F(k[x]/x^2) = k[x](-2)/x.$$

In math lingo, the left derived functors of the tensor product are called *Tor*. The reason for this is that Tor_1 represents torsion – for example, if we have an arbitrary domain R, an R-module M, and $r \in R$, then to compute $Tor_1(R/r, M)$ we take a free resolution of R/r:

$$0 \longrightarrow R \xrightarrow{\cdot r} R \longrightarrow R/r \longrightarrow 0,$$

drop the R/r and tensor with M, obtaining the complex:

$$0 \longrightarrow M \xrightarrow{\cdot r} M \longrightarrow 0.$$

So $Tor_0(R/r, M) \simeq M/rM$ (compare to Atiyah–Macdonald Exercise 2.1), and $Tor_1(R/r, M)$ is the kernel of the map from M to M defined by

$$m \longrightarrow rm.$$

$Tor_1(R/r, M)$ consists of the r-torsion elements of M, that is, the elements of M annihilated by r.

```
i1 : R=ZZ/101[x];

i2 : M=coker matrix{{x}};

i3 : N=coker matrix{{x^2}};

i4 : prune Tor_0(N,M)

o4 = cokernel | x |
```

```
                                    1
o4 : R-module, quotient of R

i5 : prune Tor_1(N,M)

o5 = cokernel {2}  | x |
                                    1
o5 : R-module, quotient of R
```

In English, $Tor_1(N, M)$ is just $R(-2)/x$, as expected. One of the most important facts about *Tor* is that we can compute $Tor_i(N, M)$ either by taking a projective resolution for N and tensoring with M, or by taking a projective resolution for M and tensoring with N. In the last section we'll prove this, but for the moment let's accept it. When I learned about derived functors and *Tor*, the first question I asked was "Why? What good is this thing?" Well, here's the beef:

Exercise 8.2.4. (Hilbert Syzygy Theorem). Let M be a finitely generated, graded module over $R = k[x_1, \ldots, x_n]$. Prove the Hilbert syzygy theorem as follows: First, compute $Tor_i(k, M)$ using a free resolution for M and tensoring with k. Now compute using a free resolution for k (remember, this is the Koszul complex from Chapter 3), and tensoring with M. Comparing the results yields the theorem, basically with no work at all! This is the power of the machinery, and it is pretty persuasive. ◇

Notice that in the betti diagram for a graded module N, the number appearing in the row indexed by j and column indexed by i is simply $dim_k Tor_i(N, k)_{i+j}$. We check the betti diagram in position $(1, 1)$ for the module N in the previous example:

```
i6 : betti N

o6 = relations : total: 1 1
                     0: 1 .
                     1: . 1

i7: hilbertFunction(2,Tor_1(N,coker vars R))

o7 = 1
```

One of the most important properties of *Tor* (indeed, of derived functors in general) is that it behaves nicely on short exact sequences.

Theorem 8.2.5. *Given a module N and short exact sequence of modules,*

$$0 \longrightarrow M_1 \longrightarrow M_2 \longrightarrow M_3 \longrightarrow 0,$$

there is a long exact sequence in Tor

$$\cdots \longrightarrow Tor_{i+1}(N, M_3) \longrightarrow Tor_i(N, M_1) \longrightarrow Tor_i(N, M_2)$$
$$\longrightarrow Tor_i(N, M_3) \longrightarrow Tor_{i-1}(N, M_1) \longrightarrow \cdots$$

Proof. Take free resolutions $(F'_\bullet, d_\bullet)$ for M_1 and $(G'_\bullet, \delta_\bullet)$ for M_3. We can obtain a free resolution for M_2 where the ith module is $F'_i \oplus G'_i$. This often called the Horseshoe lemma–work it out now! Be careful–the differential is *not* simply $d_\bullet \oplus \delta_\bullet$. So we have a short exact sequence of complexes:

$$
\begin{array}{ccccccc}
& 0 & & 0 & & 0 & \\
& \downarrow & & \downarrow & & \downarrow & \\
\cdots \longrightarrow & F'_2 & \longrightarrow & F'_1 & \longrightarrow & F'_0 & \longrightarrow 0 \\
& \downarrow & & \downarrow & & \downarrow & \\
\cdots \longrightarrow & F'_2 \oplus G'_2 & \longrightarrow & F'_1 \oplus G'_1 & \longrightarrow & F'_0 \oplus G'_0 & \longrightarrow 0 \\
& \downarrow & & \downarrow & & \downarrow & \\
\cdots \longrightarrow & G'_2 & \longrightarrow & G'_1 & \longrightarrow & G'_0 & \longrightarrow 0 \\
& \downarrow & & \downarrow & & \downarrow & \\
& 0 & & 0 & & 0 &
\end{array}
$$

Now, since the columns are short exact sequences of free modules, when we tensor this complex with N, the columns are still exact. To be completely explicit, a column is a short exact sequence of the form

$$0 \longrightarrow F'_i \longrightarrow F'_i \oplus G'_i \longrightarrow G'_i \longrightarrow 0,$$

where the first map is the identity on F'_i and zero elsewhere, and the second map is the identity on G'_i and zero on F'_i (this is part of the Horseshoe lemma). So when we tensor through with N, the first map is the identity map from rank F'_i copies of N to itself, and the second map is the identity map from rank G'_i copies of N to itself. In short, after tensoring through with N, we still have a short exact sequence of complexes, which gives a long exact sequence in homology. But these homology modules are exactly the *Tor* modules. $\square$

We close with some famous words which show that those who don't take the time to study Tor end up feeling foolish (Don't be a Tor, learn Tor!):

> "Habe nun, ach! Philosophie,
> Juristerei und Medizin,
> Und leider auch Theologie!
> Durchaus studiert, mit heissem Bemühn.
> Da steh ich nun, ich armer Tor!
> Und bin so klug als wie zuvor."
> – *Göthe, Faust, act I, scene I*

8.3 Ext

Recall that the functor $G = Hom_R(M, \bullet)$ is left exact and covariant. For a left exact, covariant functor $G : \mathcal{C} \longrightarrow \mathcal{C}$ and R–module N, we define the right derived functors $R^i G(N)$ by taking an *injective* resolution for N, applying G to the complex, and computing homology. Injective modules are defined in a fashion similar to the way we defined projective modules, but the arrows just go in opposite directions; a good reference for injective modules is Section IV.3 of [56]. Warning! Although free modules are projective, they are not in general injective. For many rings, injective modules are huge objects, good for theory but not so good for computation. What about the functor $G' = Hom_R(\bullet, M)$? It is left exact, but contravariant. For such a functor and R–module N, we define the right derived functors of $G'(N)$ by taking a projective resolution for N, applying G' to the complex, and computing homology. For the functor *Hom*, the higher derived functors have a special name: Ext^i. We continue working in the category of graded modules and graded homomorphisms.

Example 8.3.1. Let $R = k[x, y]$ and $N = R/\langle x^2, xy \rangle$; we compute $Ext^i(N, R)$. First, we take a free resolution for N:

$$0 \longrightarrow R(-3) \xrightarrow{\begin{bmatrix} x \\ -y \end{bmatrix}} R(-2)^2 \xrightarrow{[xy, x^2]} R \longrightarrow N \longrightarrow 0.$$

We drop N, and apply $Hom_R(\bullet, R)$, yielding a complex:

$$0 \longleftarrow R(3) \xleftarrow{[x, -y]} R(2)^2 \xleftarrow{\begin{bmatrix} xy \\ x^2 \end{bmatrix}} R \longleftarrow 0.$$

Since the rightmost map obviously has trivial kernel, $Ext^0(N, R) = 0$. Just as the zeroth left derived functor of a right exact, covariant functor F and module N was simply $F(N)$, the zeroth right derived functor of a left exact, contravariant functor G and module N is $G(N)$. In sum, $Ext^0(N, R) = Hom_R(N, R)$. Since every element of N is annihilated by x^2, $Hom_R(N, R)$ should indeed be zero. Then we have

$$Ext^1(N, R) = ker\,[x, -y]\,/\,im \begin{bmatrix} xy \\ x^2 \end{bmatrix} = im \begin{bmatrix} y \\ x \end{bmatrix} / im \begin{bmatrix} xy \\ x^2 \end{bmatrix} \simeq R(1)/\langle x \rangle.$$

Recall that $R(2)$ is a free module with generator of degree -2, so the generator

$$\begin{bmatrix} y \\ x \end{bmatrix}$$

of $Ext^1(N, R)$ is of degree -1. Finally, $Ext^2(N, R)$ is clearly $R(3)/\langle x, y \rangle$. We double check our computations:

```
i9  :  R=ZZ/101[x,y];

i10 :  N = coker matrix {{x^2,x*y}};

i11 :  Ext^0(N,R)

o11 =  image 0

                              1
o11 :  R - module, submodule of R

i12 :  Ext^1(N,R)

o12 =  cokernel {-1} | x |

                              1
o12 :  R - module, quotient of R

i13 :  Ext^2(N,R)

o13 =  cokernel {-3} | y x |
```

Notice that for this example the annihilators of the *Ext* modules are very easy to compute:

$$ann(Ext^0(N, R)) \simeq \langle 1 \rangle$$
$$ann(Ext^1(N, R)) \simeq \langle x \rangle$$
$$ann(Ext^2(N, R)) \simeq \langle x, y \rangle$$

For a finitely generated module $M \neq 0$ over a Noetherian ring, $Ass(M)$ contains all primes minimal over $ann(M)$ ([28], Theorem 3.1); in Section 1.3 we saw that $\{\langle x \rangle, \langle x, y \rangle\} \subseteq Ass(N)$. This is no accident, and in Chapter 10 we'll prove:

Theorem 8.3.2. *[35] Let M be a finitely generated, graded module over a polynomial ring R, and P a prime of codimension c. Then*

$$ P \in Ass(M) \Longleftrightarrow P \in Ass(Ext^c(M, R)). $$

So the *Ext* modules can help us to find the associated primes! Let's try one more example:

```
i1 : R = ZZ/31991[x,y,z,w];

i2 : I = ideal (z*w, x*w, y*z, x*y, x^3*z - x*z^3);

o2 : Ideal of R

i3 : rI=res coker gens I

        1       5       6       2
o3 = R  <--  R  <--  R  <--  R  <--  0

        0       1       2       3       4

o3 : ChainComplex

i4 : rI.dd

              1                                       5
o4 = 0 : R  <--------------------------- R  : 1
                | xy yz xw zw x3z-xz3 |
```

```
          5                                          6
    1 : R  <------------------------------ R  : 2
             {2} | -z -w 0   0   z3  0  |
             {2} | x  0  0  -w  -x3  0  |
             {2} | 0  y  -z  0   0   z3 |
             {2} | 0  0  x   y   0  -x3 |
             {4} | 0  0  0   0   y   w  |


          6                          2
    2 : R  <--------------------- R  : 3
             {3} | w  -x2w    |
             {3} | -z x2z-z3  |
             {3} | -y x2y     |
             {3} | x  0       |
             {5} | 0  -w      |
             {5} | 0  y       |


          2
    3 : R  <----- 0 : 4
             0


o4 : ChainComplexMap

i5 : betti rI

o5 = total: 1 5 6 2
         0: 1 . . .
         1: . 4 4 1
         2: . . . .
         3: . 1 2 1


i6 : Ext^2(coker gens I, R)

o6 = cokernel {-2} | z x |

                            1
o6 : R-module, quotient of R

i7 : Ext^3(coker gens I, R)
```

```
o7 = cokernel  {-4} | 0  0  x  w  -z        y |
               {-6} | w  y  0  0  x2z-z3  0 |

                              2
o7 : R-module, quotient of R
```

The annihilator of $Ext^2(R/I, R)$ is visibly the ideal $\langle z, x \rangle$. It is a tiny bit harder to see what the annihilator of $Ext^3(R/I, R)$ is, so we ask Macaulay 2:

```
i8 : annihilator o7

                       3       3
o8 = ideal (w, y, x z - x*z )

o8 : Ideal of R

i9 : primaryDecomposition o8

o9 = {ideal (x, y, w),
      ideal (z, y, w),
      ideal (x + z, y, w),
      ideal (- x + z, y,w)}

o9 : List

i10 : hilbertPolynomial coker gens I

o10 = 4*P   + P
          0     1
```

By Theorem 8.3.2 we know $\langle x, z \rangle$ is a codimension two associated prime of I, and we also have four codimension three associated primes corresponding to points. This agrees with our computation of the Hilbert polynomial. In fact, we could just do a primaryDecomposition command, and see that $V(I)$ consists of a line and four points in $\mathbb{P}^3$. Question: is there any interesting geometry associated to this set, for example, are the betti numbers what we would expect for a random choice of a line and four points in $\mathbb{P}^3$? Let's take the ideal of a random line (generated by two random linear forms) and four ideals of random points (each generated by three random linear forms), intersect and examine the resulting free resolution:

```
i11 : J=ideal random(R^{1,1},R^1)

o11 = ideal (953x + 758y + 13z + 165w,
            363x + 556y + 405z - 695w)

o11 : Ideal of R

i12 : scan(4, i->(J=intersect(J,
        ideal(random (R^{1,1,1},R^1))))))

i13 : hilbertPolynomial coker gens J

o13 = 4*P  + P
          0    1

i14 : betti res coker gens J

o14 = total: 1 3 4 2
           0: 1 . . .
           1: . 3 . .
           2: . . 4 2
```

So, the free resolution is quite different for a random configuration. A line is a line, so what must be happening is that our four points must have some special geometry (either amongst themselves, or with respect to the line). We first examine the ideal of the four generic points; we can get at this in a number of ways (think of them now!). One way is to grab the annihilator of $Ext^3(R/J, R)$:

```
i15 : annihilator Ext^3(coker gens J, R);

o15 : Ideal of R

i16 : transpose gens o15
```

```
o16 = {-2} | z2+7851xw-4805yw+9019zw+8224w2     |
       {-2} | yz+14030xw+2636yw-15494zw+1445w2  |
       {-2} | xz-5953xw+6959yw-4072zw+1380w2    |
       {-2} | y2-5477xw+9246yw+13158zw+11228w2  |
       {-2} | xy+2049xw-472yw+15528zw-13625w2   |
       {-2} | x2-2165xw+9752yw-10641zw+8888w2   |


                 6       1
o16 : Matrix R  <--- R

i17 :   betti res coker gens o15

o17 = total: 1 6 8 3
           0: 1 . . .
           1: . 6 8 3
```

By Exercise 7.1.9 this is exactly what we expect the betti diagram for four random points in $\mathbb{P}^3$ to look like. So there must be some very special geometry associated to the first set of points. Because the equations are so simple for the first variety, we can actually write down the coordinates of the four points:

$$(1:0:0:0), (0:0:1:0), (1:0:1:0), (1:0:-1:0).$$

And now we see the geometry! These four points all lie on the line $y = w = 0$. Of course, if you were watching carefully, this was also apparent in the `primaryDecomposition` on line o9. So the ideal of the four points of $V(I)$ will be defined by the two linear equations and a quartic.

Exercise 8.3.3. Have Macaulay 2 generate (or, if you are not near a keyboard, do it by hand) a free resolution for $k[x, y, z]/\langle xyz, xy^2, x^2z, x^2y, x^3 \rangle$. The betti numbers should be:

```
     total: 1 5 6 2
         0: 1 . . .
         1: . . . .
         2: . 5 6 2
```

Compute the Ext modules by hand, then check your work in Macaulay 2. ◇

Just as Tor_1 has a non-homological meaning, there is also a very important interpretation of Ext^1:

Definition 8.3.4. *Given two modules M and N, an extension of M by N is a short exact sequence*

$$0 \longrightarrow N \longrightarrow E \longrightarrow M \longrightarrow 0.$$

Two extensions are isomorphic if there is a commutative diagram of exact sequences which is the identity on M and N; an extension is trivial if $E \simeq M \oplus N$.

Theorem 8.3.5. *There is a one to one correspondence between elements of $Ext^1(M, N)$ and extensions of M by N.*

Since an extension is a short exact sequence, if we apply the functor $Hom(M, \bullet)$ to the above exact sequence, then we get a long exact sequence of Ext modules:

$$\cdots \longrightarrow Hom(M, E) \longrightarrow Hom(M, M) \xrightarrow{\delta} Ext^1(M, N) \longrightarrow \cdots .$$

It is easy to see that if $E \simeq M \oplus N$ then the connecting map δ is zero. Given an element of $Ext^1(N, M)$, how can we build an extension?

Exercise 8.3.6. Given a short exact sequence $0 \longrightarrow R \xrightarrow{i} S \xrightarrow{\pi} M \longrightarrow 0$, and a map $R \xrightarrow{j} N$ there is an obvious short exact sequence

$$0 \longrightarrow R \xrightarrow{(j,i)} N \oplus S \longrightarrow F \longrightarrow 0,$$

where F is the cokernel of (j, i). Show that mapping F to M via $(n, s) \to \pi(s)$ yields an extension $0 \longrightarrow N \longrightarrow F \longrightarrow M \longrightarrow 0$ (Hint: Snake lemma). With this result in hand, take a free resolution $F_\bullet$ of M, and apply $Hom(\bullet, N)$. An element of $Ext^1(M, N)$ is in the kernel of the map $Hom(F_1, N) \longrightarrow Hom(F_2, N)$ so gives a map from F_1/F_2 to N. Then the theorem follows from our result and the exact sequence:

$$0 \longrightarrow F_1/F_2 \longrightarrow F_0 \longrightarrow M \longrightarrow 0.$$

If you get stuck, you can find the proof on page 722 of Griffiths–Harris. $\diamond$

Exercise 8.3.7. (**Yoneda Pairing** [28], Exercise A3.27) It turns out that there is a multiplicative structure on Ext. Prove that there is a (nonzero) map

$$Ext^i(M, N) \times Ext^j(N, P) \longrightarrow Ext^{i+j}(M, P).$$

Hint: take a free resolution $\cdots F_1 \xrightarrow{d_1} F_0 \longrightarrow M$. An element of $Ext^i(M, N)$ is a map $F_i \xrightarrow{\theta} N$ such that $\theta \circ d_{i+1} = 0$. An element of $Ext^j(N, P)$ can be defined in similar fashion. Write it down and determine the pairing. $\diamond$

8.4 Double Complexes

A (first quadrant) double complex is a diagram:

$$
\begin{array}{ccccccccc}
& & \vdots & & \vdots & & \vdots & & \\
& & \downarrow & & \downarrow & & \downarrow & & \\
0 & \longleftarrow & C_{02} & \xleftarrow{\partial_{12}} & C_{12} & \xleftarrow{\partial_{22}} & C_{22} & \longleftarrow & \cdots \\
& & {\scriptstyle \delta_{02}}\downarrow & & {\scriptstyle \delta_{12}}\downarrow & & {\scriptstyle \delta_{22}}\downarrow & & \\
0 & \longleftarrow & C_{01} & \xleftarrow{\partial_{11}} & C_{11} & \xleftarrow{\partial_{21}} & C_{21} & \longleftarrow & \cdots \\
& & {\scriptstyle \delta_{01}}\downarrow & & {\scriptstyle \delta_{11}}\downarrow & & {\scriptstyle \delta_{21}}\downarrow & & \\
0 & \longleftarrow & C_{00} & \xleftarrow{\partial_{10}} & C_{10} & \xleftarrow{\partial_{20}} & C_{20} & \longleftarrow & \cdots \\
& & \downarrow & & \downarrow & & \downarrow & & \\
& & 0 & & 0 & & 0 & &
\end{array}
$$

where the maps commute and both rows and columns are complexes. We can make this into a single complex D by defining

$$
D_n = \bigoplus_{i+j=n} C_{ij},
$$

and setting the differential to be

$$
d_n(c_{ij}) = \partial_{ij}(c_{ij}) + (-1)^n \delta_{ij}(c_{ij}).
$$

Exercise 8.4.1. Check that $d_n \cdot d_{n+1} = 0$. $\diamond$

Now suppose that the columns only have homology in the bottom position. We can make a complex $E_\bullet$ via $E_i = C_{i0}/\text{image } \delta_{i1}$. It is easy to see that D_i surjects onto E_i, so we can call the kernel of this map K_i, and we get a short exact sequence of complexes:

$$
0 \longrightarrow K_\bullet \longrightarrow D_\bullet \longrightarrow E_\bullet \longrightarrow 0,
$$

yielding a long exact sequence in homology.

Exercise 8.4.2. Prove that in fact $H_j(D) \simeq H_j(E)$, i.e. $H_j(K) = 0$, for all j. Instead of trying to get this fact from generalities, it is easiest to show that the map from D_i to E_i is both injective and surjective at the level of homology, so you need to do a diagram chase. $\diamond$

By symmetry, if we have a complex whose rows only have homology at the leftmost position we can define D as before, and a complex F analogous to E, such that $H_j(D) \simeq H_j(F)$. Combining, we have:

Theorem 8.4.3. *Suppose we have a first quadrant double complex, which has homology along the rows only at the leftmost position, and homology along the columns only at the bottom position. Then you can compute homology of D by computing either the homology of E or of F.*

So what? Well, this is exactly what proves we can compute $Tor_i(M, N)$ using either a resolution for M and tensoring with N, or vice versa. Do this carefully and convince yourself that it is true. This is really just the tip of the iceberg: what we are seeing is a special case of a *spectral sequence*, a fundamental tool in many areas of algebra and topology. But it is best left for a class devoted solely to homological algebra!

Supplemental Reading: A quick overview of derived functors appears in the appendices of Matsumura's book [64]; Eisenbud [28] has a more comprehensive treatment. For the topological roots Munkres [71] is a good source. If a real surface S is given "abstractly" as a bunch of patches U_i and change of coordinate maps $U_i \overset{\delta_{ij}}{\to} U_j$; then S is orientable if for every $p \in S$, the determinant of the Jacobian matrix of δ_{ij} is positive, but this does not really give an intuitive picture. In Sections 8.2 and 8.3 we discussed derived functors for a very specific category, but where the core ideas are apparent. The reader who wants to understand more of the theoretical underpinnings will need to study Abelian categories, δ–functors, injective modules and the Baer criterion, and other derived functors like local cohomology and sheaf cohomology. These topics are covered in [28]; you can also consult a more specialized text like [98] or [42].

Chapter 9

Curves, Sheaves, and Cohomology

In this chapter we give a quick introduction to sheaves, Čech cohomology, and divisors on curves. The first main point is that many objects, in mathematics and in life, are defined by local information–imagine a road atlas where each page shows a state and a tiny fraction of the adjacent states. If you have two different local descriptions, how can you relate them? In the road map analogy, when you switch pages, where are you on the new page? Roughly speaking, a sheaf is a collection of local data, and cohomology is the mechanism for "gluing" local information together. The second main point is that geometric objects do not necessarily live in a fixed place. They have a life of their own, and we can embed the same object in different spaces. For an algebraic curve C, it turns out that the ways in which we can map C to $\mathbb{P}^n$ are related to studying sets of points (divisors) on the curve. If the ground field is $\mathbb{C}$, the maximum principle tells us that there are no global holomorphic functions on C, so it is natural to consider meromorphic functions. Hence, we'll pick a bunch of points on the curve, and study functions on C with poles only at the points. Sheaves and cohomology enter the picture because, while it is easy to describe a meromorphic function locally, it is *hard* to get a global understanding of such things. The famous theorem of Riemann–Roch tells us "how many" such functions there are for a fixed divisor. We bring the whole discussion down to earth by relating the Riemann–Roch theorem to the Hilbert polynomial, and discuss how we can use free resolutions to compute sheaf cohomology.

Key concepts: sheaves, global sections, Čech cohomology, divisors, Riemann–Roch theorem.

9.1 Sheaves

Let X be a topological space. A *presheaf* $\mathcal{F}$ on X is a rule associating to each open set $U \subseteq X$ an algebraic object (vector space, abelian group, ring . . .):

$$U \longrightarrow \mathcal{F}(U).$$

We want some simple properties to hold: for an inclusion of open sets $U \subseteq V$ there is a homomorphism

$$\mathcal{F}(V) \xrightarrow{\rho_{VU}} \mathcal{F}(U).$$

We also require that $\mathcal{F}(\emptyset) = 0$, $\rho_{UU} = id_{\mathcal{F}(U)}$, and if $U \subseteq V \subseteq W$ then composition works as expected: $\rho_{VU} \circ \rho_{WV} = \rho_{WU}$. If we wanted to be fancy, we could say that a presheaf is a contravariant functor from a category where the objects are open sets of X and the morphisms are inclusion maps to a category of algebraic objects and homomorphisms. A *sheaf* is a presheaf which satisfies two additional properties; before discussing this we give a few examples. The paradigm of a sheaf is the set of some type of functions on X; for example:

X	$\mathcal{F}(U)$
topological space	continuous real valued functions on U.
complex manifold	holomorphic functions on U.
algebraic variety	regular functions on U.

Motivated by the examples above, we'll call ρ_{VU} the restriction map; for $s \in \mathcal{F}(V)$ we write $\rho_{VU}(s) = s|_U$. Recall that being regular is a local property: a function is regular at a point p if there is a Zariski open set U around p where we can write the function as a quotient of polynomials with denominator nonzero on U. With the examples above in mind, the properties that make a presheaf a sheaf are very natural.

Definition 9.1.1. *A presheaf $\mathcal{F}$ on X is a sheaf if for any open set $U \subseteq X$ and open cover $\{V_i\}$ of U,*

1. *whenever $s \in \mathcal{F}(U)$ satisfies $s|_{V_i} = 0$ for all i, then $s = 0$ in $\mathcal{F}(U)$.*
2. *for $v_i \in \mathcal{F}(V_i)$ and $v_j \in \mathcal{F}(V_j)$ such that $v_i|_{(V_i \cap V_j)} = v_j|_{(V_i \cap V_j)}$, then there exists $t \in \mathcal{F}(V_i \cup V_j)$ such that $t|_{V_i} = v_i$ and $t|_{V_j} = v_j$.*

In English, the first condition says that something which is locally zero is also globally zero, and the second condition says that if elements agree on an overlap, then they can be glued together. The *stalk* of a sheaf at a point is the direct limit (see Appendix A)

$$\mathcal{F}_p = \varinjlim_{p \in U} \mathcal{F}(U).$$

We obtain the stalk at a point p by taking all open neighborhoods U_i of p, and then using the restriction maps to identify elements which agree on a small

enough neighborhood. This is identical to the notion of the germ of a function –
for example, if $X \subseteq \mathbb{C}$ and $\mathcal{F}$ is the sheaf of holomorphic functions, then given
$f_1 \in \mathcal{F}(U_1)$ and $f_2 \in \mathcal{F}(U_2)$ with $p \in U_i$, $f_1 = f_2$ in $\mathcal{F}_p$ if they have the same
Taylor series expansion at p, so the stalk is the ring of convergent power
series at p. Hence, the stalk gives us very local information. A morphism of
sheaves

$$\mathcal{F} \xrightarrow{\phi} \mathcal{G}$$

is defined by giving, for all open U, maps

$$\mathcal{F}(U) \xrightarrow{\phi(U)} \mathcal{G}(U),$$

which commute with the restriction map. A sequence of sheaves and morphisms is exact when it is exact locally, that is, at the level of stalks. At the
opposite extreme from the very local information given by the stalks, we have
the *global sections* of a sheaf. These are the elements of $\mathcal{F}(X)$, so are precisely
the objects defined globally on X. One thing that can seem hard to grasp at
first is that it is possible for a sheaf to be nonzero, but for it to have no nonzero
global sections. Thus, there are no global objects, although the objects exist
locally. For example, suppose we want non-constant holomorphic functions
on a compact, complex one-dimensional manifold (curve). The maximum
principle tells us that there are no such things, even though on a small enough
open set (a disk, say) they certainly exist! Just as there are lots of different
classes of algebraic objects (vector spaces, fields, rings, modules), there are
lots of types of sheaves. For algebraic geometry, the most important sheaf
is the sheaf of regular functions $\mathcal{O}_X$ on X, closely followed by *sheaves of
modules* over $\mathcal{O}_X$. A sheaf $\mathcal{F}$ of $\mathcal{O}_X$-modules is just what you expect: on an
open set U, $\mathcal{F}(U)$ is an $\mathcal{O}_X(U)$-module. For example, if Y is a subvariety of
X, then Y is defined on an open set U by a sheaf of ideals $\mathcal{I}_Y(U) \subseteq \mathcal{O}_X(U)$;
hence $\mathcal{I}_Y$ is a $\mathcal{O}_X$-module.

We've been working primarily with graded modules over a polynomial
ring; how do these objects relate to sheaves? It turns out that any "nice"
sheaf on $\mathbb{P}^n$ "comes from" a finitely generated graded module over $R =
k[x_0, \ldots, x_n]$. What this means is that *we already know* how to study sheaves
on $\mathbb{P}^n$.

Definition 9.1.2. *Suppose we have a finitely generated, graded module M
over the polynomial ring $R = k[x_0, \ldots, x_n]$, and a homogeneous $f \in R$. Then
on the Zariski open set $U_f = V(f)^c$, define a sheaf $\mathcal{M}$ via $\mathcal{M}(U_f) = \{m/f^n,
degree\ m = degree\ f^n\}$. We call $\mathcal{M}$ the sheaf associated to M, and write
$\mathcal{M} = \tilde{M}$.*

In fact, locally things are as nice as possible: for a prime ideal p the stalk $\mathcal{M}_p$ is just the degree zero piece of the localization M_p. It is not obvious that this really defines a sheaf. But it does–see [53], Section II.5 for the many details omitted here. For a projective variety X, we can build sheaves in a similar fashion.

Definition 9.1.3. *A sheaf $\mathcal{F}$ of $\mathcal{O}_X$-modules is coherent if for every point $x \in X$ there is an affine open neighborhood U_x of x with $\mathcal{F}(U_x) \simeq M$, where M is a finitely generated module over $\mathcal{O}(U_x)$.*

Definition 9.1.4. *A sheaf $\mathcal{F}$ of $\mathcal{O}_X$-modules is locally free of rank m if for every point $x \in X$, $\mathcal{F}_x \simeq \mathcal{O}_x^m$.*

Example 9.1.5. Regular functions on $\mathbb{P}^n$: let $R = k[x_0, \ldots, x_n]$. On the open patch U_{x_0}, from the definition above we have

$$\mathcal{O}_{\mathbb{P}^n}(U_{x_0}) = (R_{x_0})_0,$$

which are quotients of the form f/x_0^n, where f is homogeneous of degree n. We also have the "twisted sheaf"

$$\mathcal{O}_{\mathbb{P}^n}(m) \simeq \widetilde{R(m)}.$$

For a coherent sheaf $\mathcal{F}$ on $\mathbb{P}^n$ we set $\mathcal{F}(m) = \mathcal{F} \otimes \mathcal{O}_{\mathbb{P}^n}(m)$ and $\Gamma_*(\mathcal{F}) = \oplus_{n \in \mathbb{Z}} \mathcal{F}(n)(\mathbb{P}^n)$. For a finitely generated graded R-module M there is a homomorphism of R-modules $M \longrightarrow \Gamma_* \widetilde{M}$, which is an isomorphism in high degree. So the correspondence between coherent sheaves on $\mathbb{P}^n$ and finitely generated graded modules over R is not quite perfect – for more, see [53] Exercise II.5.9. What if $\widetilde{F}$ is a coherent sheaf on an arbitrary projective variety $X = V(I)$? Well, since mapping X into $\mathbb{P}^n$ corresponds to a map of rings from R to R/I, this means that the sheaf $\widetilde{F}$ is associated to an R/I-module F. Since we have a map from R to R/I, we can regard F as an R-module.

9.2 Cohomology and Global Sections

As we have seen, sheaves encode local data. Oftentimes it is easy to describe what is happening locally, and hard to get a global picture. In particular, the global picture involves patching things together = *cohomology*.

Example 9.2.1. For any open $U \subseteq \mathbb{R}$, let $\mathcal{F}(U)$ denote real valued continuous functions on U. Suppose we're interested in real valued continuous functions on S^1. Cover S^1 with two open sets U_1 and U_2 each $\simeq \mathbb{R}$, which overlap (at the ends) in two little open sets, each also $\simeq \mathbb{R}$. We define a map:

$$\mathcal{F}(U_1) \oplus \mathcal{F}(U_2) \longrightarrow \mathcal{F}(U_1 \cap U_2)$$

via $(f, g) \longrightarrow f - g$. Then this map has a kernel and a cokernel. The kernel will be precisely the global, real valued continuous functions on S^1, written $\mathcal{F}(S^1)$.

The Čech complex is just a generalization of the simple example above; keep in mind that what we're doing is building a cohomology where elements of H^0 are precisely the objects defined globally. For this reason, $\mathcal{F}(X)$ is also written as $H^0(\mathcal{F})$; $H^0(\mathcal{F})$ is called the zeroth cohomology (or the global sections) of the sheaf $\mathcal{F}$. Let $\mathcal{U} = \{U_i\}$ be an open cover of X – that is, a collection of open subsets of X whose union is X. When $\mathcal{U}$ consists of a finite number of open sets, then the ith module C^i in the Čech complex is simply

$$\bigoplus_{\{j_0 < \ldots < j_i\}} \mathcal{F}(U_{j_0} \cap \cdots \cap U_{j_i}).$$

Of course, in general a cover need not be finite; in this case it is convenient to think of an element of C^i as an operator c_i which assigns to each $(i + 1)$-tuple $(j_0, \ldots, j_i)$ an element of $\mathcal{F}(U_{j_0} \cap \cdots \cap U_{j_i})$. We build a complex:

$$C^i = \prod_{\{j_0 < \ldots < j_i\}} \mathcal{F}(U_{j_0} \cap \cdots \cap U_{j_i}) \xrightarrow{d^i} C^{i+1} = \prod_{\{j_0 < \ldots < j_{i+1}\}} \mathcal{F}(U_{j_0} \cap \cdots \cap U_{j_{i+1}}),$$

where $d^i(c_i)$ is defined by how it operates on $(i + 2)$-tuples, which is:

$$d^i(c_i)(j_0, \ldots, j_{i+1}) = \sum_{k=0}^{i+1} (-1)^k c_i(j_0, \ldots, \hat{j}_k, \ldots, j_{i+1})|_{\mathcal{F}(U_{j_0} \cap \cdots \cap U_{j_{i+1}})}.$$

Now we define the Čech cohomology *of a cover* $\mathcal{U}$ as the cohomology of this complex.

Exercise 9.2.2. (**Čech cohomology with $\mathbb{Z}$-coefficients**) The constant sheaf on a space X is defined by giving $\mathbb{Z}$ the discrete topology, and then setting $\mathbb{Z}(U)$ to be continuous functions from U to $\mathbb{Z}$. (remark: there is an easy extension of this definition from $\mathbb{Z}$ to any abelian group)

1. Is it true that $\mathbb{Z}(U) \simeq \mathbb{Z}$ for any open set U? If not, what additional assumption on U would make this true?

2. Compute the Čech cohomology of the constant sheaf $\mathbb{Z}$ on S^2 using the cover of the open top hemisphere, and the two open bottom "quarter-spheres" (all opens slightly enlarged so that they overlap) and see what you get. Your chain complex should start with these three opens (each of which is topologically an $\mathbb{R}^2$), then the three intersections (each of which is again an $\mathbb{R}^2$) and then the triple intersection, which is two disjoint $\mathbb{R}^2$'s. Of course, you'll need to write down the differentials in this complex (it is not bad, and will help you see how the definition works!).

3. Next, use a hollow tetrahedron to approximate S^2, and write down the Čech complex corresponding to open sets which are (slightly enlarged so they overlap) triangles. Compare this to the exercise on simplicial cohomology in Chapter 6.

4. What happens if you use the open cover consisting of the top hemisphere and bottom hemisphere? ◇

Formally, the sheaf cohomology is defined as a direct limit, over all open covers, of the cohomology of the covers. Of course, this is no help at all for actually computing examples, but it can be shown that if all intersections of the open sets of a cover have no cohomology except H^0, then that cover actually can be used to compute the cohomology (such a cover is called a *Leray* cover). Can you see which of the covers above is a Leray cover? A very important fact is that the open cover of $\mathbb{P}^n$ given by $\{U_{x_i}\}_{i=0}^n$ is a Leray cover (see [53] III.3 for a proof).

As we have seen many times, one way to understand a module is to fit it into an exact sequence. In the situation of interest to us (when the $\mathcal{F}_i$ are sheaves of modules on a smooth algebraic variety X) we have the following key theorem:

Theorem 9.2.3. *Given a short exact sequence of sheaves* (*remember that exact means exact on stalks*)

$$0 \longrightarrow \mathcal{F}_1 \longrightarrow \mathcal{F}_2 \longrightarrow \mathcal{F}_3 \longrightarrow 0,$$

there is a long exact sequence in sheaf cohomology

$$\cdots \longrightarrow H^{i-1}(\mathcal{F}_3) \longrightarrow H^i(\mathcal{F}_1) \longrightarrow H^i(\mathcal{F}_2) \longrightarrow H^i(\mathcal{F}_3)$$
$$\longrightarrow H^{i+1}(\mathcal{F}_1) \longrightarrow \cdots .$$

The proof rests on an alternate approach to defining sheaf cohomology. First, the global section functor is left exact and covariant. Second, injective resolutions of $\mathcal{O}_X$-modules exist (in general, the existence of injective resolutions is not automatic), so the results of Chapter 8 imply that a short exact sequence of sheaves yields a long exact sequence in the higher derived functors of H^0. Showing that these higher derived functors agree with the Čech cohomology would suffice to prove the theorem. This (along with the existence of an injective resolution) requires some work, and can be found in Section III.4 of [53]. The most important instance of a short exact sequence of sheaves is the ideal sheaf sequence. For a variety X sitting inside $\mathbb{P}^n$, say $X = V(I)$, we have the good old exact sequence of modules:

$$0 \longrightarrow I \longrightarrow R \longrightarrow R/I \longrightarrow 0.$$

Since exactness is measured on stalks, if we take an exact sequence of modules and look at the associated sheaves, we always get an exact sequence of sheaves. So we have an exact sequence of sheaves on $\mathbb{P}^n$:

$$0 \longrightarrow \mathcal{I}_X \longrightarrow \mathcal{O}_{\mathbb{P}^n} \longrightarrow \mathcal{O}_X \longrightarrow 0.$$

Exercise 9.2.4. Prove that for any m, $H^i(\mathcal{O}_{\mathbb{P}^n}(m)) = 0$ unless $i = 0$ or $i = n$. Use the definition of Čech cohomology and the fact that the standard open cover of $\mathbb{P}^n$ is Leray. Hint: [53] Theorem III.5.1. $\diamond$

In particular, if $n > 1$, then

$$H^1(\mathcal{O}_{\mathbb{P}^n}(m)) = 0.$$

This means that for any projective variety $X \subseteq \mathbb{P}^n$, $n \geq 2$ we'll *always* have an exact sequence:

$$0 \longrightarrow H^0(\mathcal{I}_X(m)) \longrightarrow H^0(\mathcal{O}_{\mathbb{P}^n}(m)) \longrightarrow H^0(\mathcal{O}_X(m))$$
$$\longrightarrow H^1(\mathcal{I}_X(m)) \longrightarrow 0.$$

This is what we were computing when we dealt with points back in Chapter 7!

Exercise 9.2.5. In Section 4 we'll see that $H^0(\mathcal{O}_{\mathbb{P}^1}(3))$ is a finite dimensional vector space. Use the definition of Čech cohomology and the standard open cover of $\mathbb{P}^1$ to compute a basis of $H^0(\mathcal{O}_{\mathbb{P}^1}(3))$ (you should find 4 elements). For X a point, prove $H^0(\mathcal{O}_X)$ is one-dimensional, and all higher cohomology vanishes. $\diamond$

9.3 Divisors and Maps to $\mathbb{P}^n$

A major step in geometry came when people realized that it was important to distinguish between intrinsic properties of an object and extrinsic properties: those properties that depend on how the object is embedded in some space. An embedding of a variety X is an isomorphism between X and a variety in $\mathbb{P}^n$. A simple example of this is the projective line. First, it has a life of its own, as plain old $\mathbb{P}^1$. On the other hand, it can also be embedded in $\mathbb{P}^2$. Let a, b be homogeneous coordinates on $\mathbb{P}^1$, and x, y, z be homogeneous coordinates on $\mathbb{P}^2$; we now embed $\mathbb{P}^1$ in two different ways. First, let

$$\mathbb{P}^1 \xrightarrow{\ f\ } \mathbb{P}^2,$$

be defined by

$$(a : b) \xrightarrow{\ f\ } (a^2 : ab : b^2).$$

The image of f lies in the region of $\mathbb{P}^2$ covered by the open sets U_x and U_z. On $U_x, (x : y : z) = (1 : \frac{y}{x} : \frac{z}{x})$ which we write affinely as $(\frac{y}{x}, \frac{z}{x})$. The inverse map h from U_x to U_a which sends $(s, t) \longrightarrow s$ takes $(\frac{y}{x}, \frac{z}{x})$ to $\frac{y}{x}$, which in homogeneous terms yields

$$\left(1 : \frac{y}{x}\right) = \left(1 : \frac{ab}{a^2}\right) = \left(1 : \frac{b}{a}\right) = (a : b),$$

so on U_a, $h \circ f$ is the identity. A similar computation on U_z shows that f is an embedding. The image of f is $V(xz - y^2)$, and so the Hilbert polynomial is

$$HP(k[x, y, z]/\langle zx - y^2 \rangle, t) = 2t + 1.$$

Now let

$$\mathbb{P}^1 \xrightarrow{\ g\ } \mathbb{P}^2,$$

be defined by

$$(a : b) \xrightarrow{\ g\ } (a : b : 0).$$

This is again an embedding. But now the Hilbert polynomial of the image is

$$HP(k[x, y, z]/\langle z \rangle, t) = t + 1.$$

In particular, *the Hilbert polynomial depends upon the embedding*. Of course, we knew this already, because the lead coefficient of the Hilbert polynomial of a curve is the degree. The point is that the same object (plain old $\mathbb{P}^1$) can have different incarnations. What we are seeing is that there are two parts to "understanding" a variety – what it is intrinsically, and in what ways we can embed it in some projective space. For the remainder of this chapter we restrict to the case of smooth, complex algebraic curves. It turns out that the way to attack the second part of the problem is by studying sets of points on the curve C. First, we need some generalities:

A *compact Riemann surface* is a compact, complex manifold of dimension one (for the definition of manifold, see [49], the intuitive idea is that a complex one-dimensional manifold "locally looks like" $\mathbb{C}$). The word surface appears because there are two real dimensions; topologically these gadgets are the objects X_g which appeared in Exercise 8.1.7. Recall that the *genus g* is just the number of holes in X_g, so $2g$ is the rank of the first simplicial homology of a triangulation of X_g. The key fact is that any compact Riemann surface is a smooth algebraic curve. This is pretty amazing – why should there be any algebraic structure at all? The relevance of these facts to the problem of mapping a curve to projective space is that we can employ tools of complex analysis to study the problem. Since there are no global holomorphic functions on a curve, it is natural to consider meromorphic functions: pick a bunch of points on the curve, and consider functions on C with poles only at the points. This is the essential idea.

In the last section we touched on the notion of an ideal sheaf. Since a curve is one-dimensional, if we look at the ideal of a point (on some open patch), the ideal will be principal. Now, of all the ideals in the world, only one type of ideal corresponds to a locally free sheaf – those which are locally principal! The second thing to notice is that if an ideal sheaf is defined locally by a polynomial f, and if we want to consider meromorphic functions with poles only at the points where f vanishes, then locally this sheaf is generated by $1/f$. Pick a (finite) set of points p_i where poles will be allowed (possibly with order greater than one, but always finite).

Definition 9.3.1. *A divisor on a curve C is a finite integral combination of points of C:*

$$D = \sum a_i p_i.$$

The degree of D is $\sum a_i$.

To a divisor D we associate a locally free sheaf, usually written as $\mathcal{O}(D)$ or $\mathcal{L}(D)$. To do this, take a cover of C by open sets, such that at each point p_i of D, there is a little open disk U_i with local coordinate z_i, centered at p_i (i.e. $z_i(p_i) = 0$). For a meromorphic function f defined on U_i, a power series expansion of f will have the form

$$f(z_i) = z_i^n (c_0 + c_1 z_i + c_2 z_i^2 + \cdots), c_0 \neq 0.$$

If $n > 0$ then f has a zero of order n at p_i, written $n \cdot p_i \in Z(f)$; if $n < 0$ then f has a pole of order n at p_i, written $-n \cdot p_i \in P(f)$. Now set

$$div(f) = Z(f) - P(f).$$

Definition 9.3.2. *The sheaf* $\mathcal{O}(D)$ *consists on an open set* U *of meromorphic functions* f *on* U *such that*

$$div(f) + D|_U \geq 0.$$

So if $a_i > 0$ this means we're allowing a pole at p_i of order at most a_i, and if $a_i < 0$ this means we require a zero of order at least a_i. If we can find *global* sections of $\mathcal{O}(D)$, then we'll have a chance of mapping C to some projective space, by using the global sections (which are meromorphic functions) as our map. In the next section we'll prove that $H^0(\mathcal{O}(D))$ is a finite dimensional complex vector space; the Riemann–Roch problem is to determine $dim_\mathbb{C} H^0(\mathcal{O}(D))$. If $\{f_0, \ldots, f_k\}$ is a basis for $H^0(\mathcal{O}(D))$, then we obtain a map

$$C \xrightarrow{(f_0, \ldots, f_k)} \mathbb{P}^k.$$

D is said to be *very ample* if $H^0(\mathcal{O}(D))$ defines an embedding; if $deg(D) > 2g$, then ([53], IV.3) D is very ample.

Definition 9.3.3. *Two divisors* D_1, D_2 *are linearly equivalent* ($D_1 \simeq D_2$) *if they differ by the divisor of a meromorphic function.*

Example 9.3.4. On $\mathbb{P}^1$ let's consider the divisor $D = 3p$, where p is the point $(1 : 0)$. Write $k[X, Y]$ for the coordinate ring of $\mathbb{P}^1$, and let x denote X/Y and y denote Y/X. On U_Y the ideal sheaf of D is 1 since the support of D does not meet U_Y. On U_X the ideal sheaf of D is generated by y^3. A point given globally as $(p_0 : p_1)$ is written locally on U_Y as p_0/p_1, and when we change from U_Y to U_X, it is transformed to p_1/p_0, so on $U_Y \cap U_X, x = 1/y$. We have that

$$\mathcal{O}(D)(U_Y) = 1 \cdot \mathcal{O}(U_Y), \mathcal{O}(D)(U_X) = 1/y^3 \cdot \mathcal{O}(U_X).$$

Thus, $H^0(\mathcal{O}(D))$ has a basis consisting of the following pairs of elements of $(\mathcal{O}(D)(U_Y), \mathcal{O}(D)(U_X))$:

$$(x^3, 1), (x^2, y), (x, y^2), (1, y^3).$$

We can think of these as elements of the coordinate ring

$$X^3, X^2Y, XY^2, Y^3.$$

Warning: although a homogeneous polynomial has a well defined zero set on projective space, it is *NOT* a function on projective space. But taken together, the global sections give us a map from $\mathbb{P}^1$ to $\mathbb{P}^3$ by sending

$$(a:b) \longrightarrow (a^3 : a^2b : ab^2 : b^3).$$

The image of this map is our old friend the twisted cubic. Notice that if we slice the image with a hyperplane, we get three points. This is because a generic hyperplane of $\mathbb{P}^3$ corresponds exactly to the vanishing of a generic cubic on $\mathbb{P}^1$.

```
i1 : R=ZZ/31991[x,y];

i2 : S=ZZ/31991[a,b,c,d];

i3 : map(R,S,matrix{{x^3,x^2*y,x*y^2,y^3}})

                  3   2     2   3
o3 = map(R,S,{x , x y, x*y , y })

o3 : RingMap R <--- S

i4 : kernel o3

              2                        2
o4 = ideal (c  - b*d, b*c - a*d, b  - a*c)

o4 : Ideal of S

i5 : betti res coker gens o4

o5 = total: 1 3 2
         0: 1 . .
         1: . 3 2
i6 : hilbertPolynomial coker gens o4

o6 = - 2*P   + 3*P
          0       1
```

Exercise 9.3.5. Consider the cubic curve $C = V(X^3 + Y^3 + Z^3) \subseteq \mathbb{P}^2$. We pick a divisor D of degree six and use the global sections as a map. An easy choice is to take the intersection of C with the curve $V(XY)$, so that D consists of two sets of three collinear points. Check that on U_Z

$$\left\{ \frac{x}{y}, 1, \frac{1}{y}, \frac{y}{x}, \frac{1}{x}, \frac{1}{xy} \right\}$$

are sections, and as in the previous example we can write these globally as

$$\{X^2, XY, XZ, Y^2, YZ, Z^2\}.$$

It is not obvious that these are all the sections, but we'll see that this is so in the next section. The observant reader may have noticed that the monomials we wrote down above are sections of $\mathcal{O}_{\mathbb{P}^2}(2)$, so we actually have a map ϕ from all of $\mathbb{P}^2$ to $\mathbb{P}^5$. The image of $\mathbb{P}^2$ is called the *Veronese surface*; ϕ carries the curve C along for the ride. First part: Let $(a_0 : \ldots : a_5)$ be coordinates on $\mathbb{P}^5$; show that the Veronese surface is defined by the vanishing of the two by two minors of

$$\begin{bmatrix} a_0 & a_1 & a_2 \\ a_1 & a_3 & a_4 \\ a_2 & a_4 & a_5 \end{bmatrix}.$$

Second part: Since $X^3 + Y^3 + Z^3 = 0$, so is $X(X^3 + Y^3 + Z^3)$, which we can rewrite as $a_0^2 + a_1 a_3 + a_2 a_5 = 0$. Find similar relations and determine the equations for the image of C in $\mathbb{P}^5$. Finally, with the equations in hand, have Macaulay 2 tell you the Hilbert polynomial, and see if we really get a curve of degree six. A generic hyperplane in $\mathbb{P}^5$ corresponds to a general quadric in $\mathbb{P}^2$, which meets C in *six* points.

```
i2 : R=ZZ/101[a_0..a_5];

i3 : S=ZZ/101[X,Y,Z];

i4 : m = basis(2,S)

o4 = | X2 XY XZ Y2 YZ Z2 |

                 1       6
o4 : Matrix S  <--- S

i5 : map(S,R,m)
```

```
                      2              2          2
o5 = map(S,R,{X , X*Y, X*Z, Y , Y*Z, Z })

o5 : RingMap S <--- R

i7 : transpose gens ker o5

o7 = {-2} | a_4^2-a_3a_5  |
     {-2} | a_2a_4-a_1a_5 |
     {-2} | a_2a_3-a_1a_4 |
     {-2} | a_2^2-a_0a_5  |
     {-2} | a_1a_2-a_0a_4 |
     {-2} | a_1^2-a_0a_3  |
--are these really the minors?

i8 : n=matrix{{a_0,a_1,a_2},{a_1,a_3,a_4},
              {a_2,a_4,a_5}}

o8 = | a_0 a_1 a_2 |
     | a_1 a_3 a_4 |
     | a_2 a_4 a_5 |

i9 : minors(2,n) == ker o5

o9 = true

i10 : p=ideal(a_0^2+a_1*a_3+a_2*a_5,
              a_0*a_1+a_3^2+a_4*a_5,
              a_0*a_2+a_3*a_4+a_5^2);

o10 = Ideal of R

i11 : q = ((ker o5)+ p);

o11 : Ideal of R

i12 : betti res coker gens q

o12 = total: 1 9 16 9 1
          0: 1 .  .  . .
          1: . 9 16 9 .
          2: . .  .  . 1
```

```
i13 : hilbertPolynomial coker gens q
```

```
o13 = - 6*P   + 6*P
           0       1  ◇
```

9.4 Riemann–Roch and Hilbert Polynomial Redux

The cornerstone of the theory of algebraic curves is the Riemann–Roch theorem, which answers the question we studied last section: How many global sections does $\mathcal{O}(D)$ have? We start by proving that $H^0(\mathcal{O}(D))$ is a finite-dimensional complex vector space; we write $h^0(\mathcal{O}(D))$ for $\dim_{\mathbb{C}} H^0(\mathcal{O}(D))$.

Definition 9.4.1. *A holomorphic (meromorphic) one-form ω is defined on a local patch U_1 with coordinate z_1 as $g_1(z_1)dz_1$, where $g_1(z_1)$ is holomorphic (meromorphic). On a patch U_2 with coordinate z_2 such that $U_1 \cap U_2$ is nonempty, $g_1(z_1)dz_1$ transforms into $g_2(z_2)dz_2$ via the chain rule. The divisor of ω is the divisor defined by the $g_i(z_i)$.*

If ω_1 and ω_2 are holomorphic one-forms, then the ratio of their coefficient functions is meromorphic and we obtain a well defined divisor class K, called the *canonical* divisor. On a curve C, the sheaf of holomorphic one-forms Ω_C^1 corresponds to the sheaf $\mathcal{O}(K)$. The vector space of global holomorphic one-forms has $h^0(\Omega_C^1) = g$, where g is the genus of the underlying real surface. This is non-trivial, for a proof, see [48].

Example 9.4.2. On $\mathbb{P}^1$ we take the one-form dy on U_X. Then since we send (p_1/p_0) on U_X to (p_0/p_1) on U_Y, the map sends a local coordinate y to $1/y = x$. So $dx = -dy/y^2$, i.e. we get a pole of order two, and the canonical divisor on $\mathbb{P}^1$ is -2 times a point.

Exercise 9.4.3. Prove that on $\mathbb{P}^1$ any two points are linearly equivalent.

To prove that $h^0(\mathcal{O}(D))$ is finite, consider a divisor D on C. For simplicity, let's assume that the points of D are distinct, so $D = \sum_{i=1}^m p_i$. Then $h^0(\mathcal{O}(D))$ is bounded by $deg(D) + 1$: we map an element of $H^0(\mathcal{O}(D))$ to its residues at the points of D (if your complex analysis is rusty and you are thinking "What fresh hell is this?"[1], you may want to cut to Appendix B for a refresher). This gives us a map to $\mathbb{C}^{deg(D)}$, and obviously the constant functions are in the kernel

---

[1] Dorothy Parker

of the map; in fact any two elements which have the same principal part are equal up to constants, since their difference is holomorphic. So $h^0(\mathcal{O}(D)) \leq deg(D) + 1$. Next we study the cokernel of the map $H^0(\mathcal{O}(D)) \to \mathbb{C}^{deg(D)}$. Let D_i be little disks centered at the p_i, with $\partial D_i = \gamma_{p_i}$, and put $W = C - \cup D_i$. If we have a meromorphic one-form α on C with simple poles at the p_i, then

$$\sum_{p \in C} \text{res}_p \alpha = \sum_{p_i} \frac{1}{2\pi i} \int_{\gamma_{p_i}} \alpha = -\frac{1}{2\pi i} \int_{\partial W} \alpha = 0,$$

since α is holomorphic on W. Thus, if we take $f \in H^0(\mathcal{O}(D))$ and $\omega \in H^0(\Omega^1)$ we have that

$$\sum_{p \in C} \text{res}_p(f \cdot \omega) = 0.$$

In English, each holomorphic one-form imposes a condition on residues at the points of D, unless the one-form vanishes at D. Thus, we get $g - h^0(\mathcal{O}(K - D))$ conditions. It can be shown that they are independent, and so:

$$h^0(\mathcal{O}(D)) \leq deg(D) + 1 - g + h^0(\mathcal{O}(K - D));$$

in fact, equality holds:

Theorem 9.4.4 (Riemann–Roch).

$$h^0(\mathcal{O}(D)) = deg(D) + 1 - g + h^0(\mathcal{O}(K - D)).$$

Now that we have at least sketched Riemann–Roch, let's see what it has to say about things we already know. For a curve C embedded in $\mathbb{P}^n$ with $C = V(I)$, we know we have a free resolution of the form:

$$0 \longrightarrow F_{n+1} \longrightarrow F_n \longrightarrow \cdots \longrightarrow R \longrightarrow R/I \longrightarrow 0.$$

This sequence remains exact when we pass to associated sheaves, since it is exact locally. We have $\mathcal{O}_C = \widetilde{R/I}$, and the alternating sum of the dimensions of the cohomology modules (usually called the sheaf-theoretic Euler characteristic)

$$\chi(\mathcal{O}_C) = \sum (-1)^i h^i(\mathcal{O}_C)$$

may be computed as the alternating sum $\sum (-1)^i \chi(\widetilde{F}_i)$. If a basis $\{f_0, \ldots, f_n\}$ for $H^0(\mathcal{O}(D))$ is used to map $C \longrightarrow \mathbb{P}^n$, then a hyperplane $V(\sum_{i=0}^n a_i x_i) \subseteq \mathbb{P}^n$ meets C when $\sum_{i=0}^n a_i f_i = 0$. We can use Riemann-Roch to compute the Hilbert polynomial of R/I_C: the F_i are free modules, and [53] Theorem III.5.1 tells us that

$$H^i(\mathcal{O}_{\mathbb{P}^n}(t)) = 0, i \geq 1, t \geq 0 \text{ and } H^0(\mathcal{O}_{\mathbb{P}^n}(t)) = R_t.$$

So it follows that

$$\chi(\mathcal{O}_C(tD)) = HP(R/I, t).$$

The Hilbert polynomial has reappeared! In sum, the Hilbert polynomial which we defined in Chapter 2 is indeed the same thing as the Hilbert polynomial you'll find in Section III.5 of [53].

Exercise 9.4.5. In any of the books on curves listed in the references, you can find a proof of *Serre duality*, which for a divisor D on a curve says

$$dim_{\mathbb{C}} H^1(\mathcal{O}(D)) = h^0(\mathcal{O}(K - D)).$$

Now assume D is *effective*: $D \simeq \sum a_i p_i$ with $a_i \geq 0$. Use Serre duality to give another proof of Riemann–Roch as follows: If D is the divisor used to embed C in $\mathbb{P}^n$, then we have an exact sequence

$$0 \longrightarrow \mathcal{I}_{t \cdot D} \longrightarrow \mathcal{O}_C \longrightarrow \mathcal{O}_{t \cdot D} \longrightarrow 0.$$

1. Find an appropriate object to tensor with so that the above sequence becomes:

$$0 \longrightarrow \mathcal{O}_C \longrightarrow \mathcal{O}_C(t \cdot D) \longrightarrow \mathcal{O}_{t \cdot D} \longrightarrow 0.$$

This shows that

$$\chi(\mathcal{O}_C(t \cdot D)) = \chi(\mathcal{O}_C) + \chi(\mathcal{O}_{t \cdot D}).$$

2. By definition $\chi(\mathcal{O}_C) = h^0(\mathcal{O}_C) - h^1(\mathcal{O}_C)$. Combine Serre duality (with D the empty divisor), the fact that $h^0(\mathcal{O}_C(K)) = g$ and the exercise at the end of the last section (that $\chi(\mathcal{O}_{point}) = 1$) to obtain:

$$\chi(\mathcal{O}_C(t \cdot D)) = 1 - g + deg(D) \cdot t. \quad \diamond$$

Example 9.4.6. For Example 9.3.4, we know that $\mathbb{P}^1$ has genus zero, since the corresponding real manifold is S^2. We also know that the canonical divisor on $\mathbb{P}^1$ is -2 times a point. Thus, $\mathcal{O}_{\mathbb{P}^1}(K - tD)$ will have no global sections, since there are no meromorphic functions with $3t + 2$ zeroes and no poles. Riemann–Roch then yields that

$$h^0(\mathcal{O}_C(t \cdot D)) = 3t + 1,$$

which is the Hilbert polynomial of the twisted cubic. For Exercise 9.3.5, we use the fact that a smooth plane curve C of degree d has genus

$$g = \binom{d-1}{2}.$$

To prove this, let $R = k[x, y, z]$, take a free resolution and pass to sheaves:

$$0 \longrightarrow R(-d+t) \longrightarrow R(t) \longrightarrow R(t)/I_C \longrightarrow 0,$$
$$0 \longrightarrow \mathcal{O}_{\mathbb{P}^2}(t-d) \longrightarrow \mathcal{O}_{\mathbb{P}^2}(t) \longrightarrow \mathcal{O}_C(t) \longrightarrow 0.$$

We obtain

$$1 - g + deg(D) \cdot t = \chi(\mathcal{O}_C(t \cdot D)) = HP(R/I_C, t)$$
$$= \binom{t+2}{2} - \binom{t-d+2}{2} = dt + 1 - \binom{d-1}{2}.$$

In particular, a smooth plane cubic has genus one. From Riemann–Roch and the fact that $h^0(\mathcal{O}(K)) = g$, it follows that degree $K = 2g - 2$, so K has degree zero on a genus one curve. For a divisor of degree six, we find $h^0(\mathcal{O}_C(K - tD)) = 0$ for t positive, so Riemann–Roch gives another reason that the Hilbert polynomial in Exercise 9.3.5 is $6t$.

Exercise 9.4.7. Suppose D is an effective divisor on a curve C, and let ϕ_K be the map defined by the canonical divisor $C \xrightarrow{\phi_K} \mathbb{P}^{g-1}$. Let $\overline{D}$ be the linear subspace of $\mathbb{P}^{g-1}$ spanned by $\phi_K(D)$. Prove the *Geometric* Riemann–Roch theorem: $h^0(\mathcal{O}(D)) = deg(D) - \dim \overline{D}$. Hint: relate the codimension of $\overline{D}$ to the space $H^0(\mathcal{O}(K - D))$ of holomorphic differentials which vanish on D. ◇

Example 9.4.8. As in Exercise 9.3.5, take a smooth cubic $C \subseteq \mathbb{P}^2$, but this time take a divisor D of degree five. Since a divisor of negative degree has no global sections (since global holomorphic functions on C are constants, there are certainly no global holomorphic functions which in addition have prescribed zeroes), $h^0(\mathcal{O}(K - D)) = 0$ (we say that D is *nonspecial*). Riemann–Roch shows that $h^0(\mathcal{O}(D)) = 5 + 1 - 1 = 5$, hence we obtain a map of C to $\mathbb{P}^4$. To find $HF(R/I_C, t)$, we just use Riemann–Roch to compute $\chi(\mathcal{O}_C(tD))$. Let's see how many quadrics are in I_C: by Riemann–Roch,

$$HF(R/I_C, 2) = \chi(\mathcal{O}_C(2D)) = deg(2D) + 1 - g = 10.$$

But in $\mathbb{P}^4$, there are $\binom{4+2}{2} = 15$ quadrics. Hence, I_C contains 5 quadrics. Let's keep going: $HF(R/I_C, 3) = 15$, whereas there are $\binom{4+3}{3} = 35$ cubics in five variables. Each quadric spawns five cubics (multiply by the variables), so there are 25 cubics in I_C which come from quadrics. Since there are only 20 cubics in I_C, there must be *at least* five linear dependencies, i.e. there are at least five linear syzygies on the quadrics.

One can go on in this way. We say that a map is given by a *complete linear system* if the map is defined by using all the global sections of some divisor. If C is embedded by a complete linear system corresponding to a nonspecial divisor D, Castelnuovo's base point free pencil trick ([28], Exercise 17.18) shows that $H^1(\mathcal{I}_C(m)) = 0$ if $m \neq 2$. If D is a very ample divisor such that $H^1(\mathcal{I}_C(m)) = 0$ for all m, then the homogeneous coordinate ring R/I_C is called *projectively normal* or *arithmetically Cohen–Macaulay* (fear not, this is explained in the next chapter). This means that the length of a minimal free resolution of R/I_C is equal to the codimension of C, which for the example above is obviously three. It turns out that there are exactly five linear syzygies, and I_C is closely related to this five-by-five matrix of linear forms.

Exercise 9.4.9. Pick a divisor of degree five on a smooth plane cubic curve and check the claims above in Macaulay 2. Once you have the matrix of linear forms, try the Macaulay 2 command `pfaffians` and see what you get. For more on this example, see Buchsbaum and Eisenbud [22]. Now try mimicking the argument above for D of degree six, and check your result against the resolution that appeared in Exercise 9.3.5. ◇

We close by returning to the central point of this chapter. A curve C does not live in a projective space until we choose an embedding, which depends on a divisor D. Once we choose D and use the sections $H^0(\mathcal{O}_C(D)) = V$ to map $C \longrightarrow \mathbb{P}(V)$, *then* it makes sense to study the free resolution of R/I_C. So the big question is: how does this free resolution relate to the choice of D? Here's an example of a famous question along these lines: What does the free resolution of R/I_C look like when C is embedded by the canonical divisor K? For a curve of genus greater than two which is not hyperelliptic (which means C does not possess a divisor D with $deg(D) = 2$ and $h^0(\mathcal{O}(D)) = 2$; in particular C does not admit a $2:1$ map to $\mathbb{P}^1$), K gives an embedding; Mark Green has conjectured that if C is sufficiently general in the family of curves of fixed genus, then there are no syzygies except linear ones for the first $\lfloor \frac{g-3}{2} \rfloor$ steps. For substantial recent progress on this, see Teixidor [94] and Voisin [96].

Exercise 9.4.10. Do the Macaulay 2 tutorial on Divisors, and then try the tutorial on canonical embeddings of plane curves and gonality. These are super exercises and also lots of fun! ◇

Mea maxima culpa: In this chapter we have sketched material more properly studied in a semester course. We have either simply stated or at best tersely sketched major theorems; we moved without justification between regular and holomorphic functions and ignored important facts (e.g. for a coherent sheaf $\mathcal{F}$ on a projective variety defined over a field k, $H^i(\mathcal{F})$ is a finite dimensional k vector space). All of these failings can be remedied by digging into one of the texts below. For those with a course in one complex variable (which we have assumed), the little book of Griffiths [48] is an excellent place to start.

Supplemental Reading: For those who want a more algebraic approach, Fulton [40] is nice. However, just as the fundamental theorem of algebra is best proved using complex analysis, I think that the complex analytic viewpoint is the proper way to approach the subject for the first time. Miranda [69] is very comprehensive, encompasses both viewpoints, and is a nice read. Hartshorne's chapters on cohomology and on curves are also very good; in fact, it probably makes sense for a first pass at Hartshorne to start at Chapter 4. Other references for curves are [2], [18], [33], [74], [97]. See pages 166–167 of Harris [52] for more on the Hilbert polynomial and Riemann–Roch. In Chapter 8 of [95], Eisenbud gives a beautiful explanation of how to compute sheaf cohomology. A closely related question concerns the computation of sheaf-theoretic versions of Ext; for this, see Smith [85].

Chapter 10

Projective Dimension, Cohen–Macaulay Modules, Upper Bound Theorem

Since free modules are projective, a module M always has a projective (possibly infinite) resolution. The minimum length of a projective resolution is the *projective dimension* of M, written $pdim(M)$. When M is a finitely generated, graded module over $k[x_1, \ldots, x_n]$ then we saw in Chapter 3 that $pdim(M)$ is just the length of a minimal *free* resolution. In this chapter we examine the relation between $pdim(M)$ and the geometry of M. When M is finitely generated over a graded or local ring R with maximal ideal $\mathfrak{m}$, we can characterize $pdim(M)$ as

$$sup\{i \,|\, Tor_i(M, R/\mathfrak{m}) \neq 0\}.$$

An important invariant of a module M is its *depth*, which we define and relate to the associated primes of M. The famous Auslander–Buchsbaum theorem gives a precise relationship between $depth(M)$ and $pdim(M)$.

We then investigate Cohen–Macaulay modules, which are modules where there is a particularly close connection between the projective dimension and the geometry of the associated primes. We briefly revisit the material of Chapter 9, discussing the relationship between the linear system used to map a curve C to $\mathbb{P}_k^n$ and the Cohen–Macaulay property for the $R = k[x_0, \ldots, x_n]$-module R/I_C. Finally, we tie everything up by sketching Stanley's wonderful proof of the upper bound conjecture for simplicial spheres, where Cohen–Macaulay modules play a leading role.

Key concepts: projective dimension, Krull dimension, Cohen–Macaulay, upper bound theorem for simplicial spheres.

10.1 Codimension, Depth, Auslander–Buchsbaum Theorem

In Chapter 3, we defined the codimension of a homogeneous ideal $I \subseteq R = k[x_0, \ldots, x_n]$ as n minus the degree of the Hilbert polynomial of R/I, and showed that this was sensible from a geometric standpoint. While this definition had the virtue of being easy to grasp and easy to compute, it does not make sense for other rings. However, it is compatible with the following general definition:

Definition 10.1.1. *For a prime ideal P, the codimension of P is the supremum of chains of prime ideals contained in P. For arbitrary I, the codimension of I is the smallest codimension of a prime containing I.*

Example 10.1.2. For $I = \langle x, y, z \rangle \subseteq k[x, y, z]$, we have a chain of primes

$$0 \subsetneq \langle x \rangle \subsetneq \langle x, y \rangle \subsetneq \langle x, y, z \rangle,$$

so I has codimension at least three (we count the number of strict inclusions). For $I = \langle xy, xz \rangle \subseteq k[x, y, z]$, $\langle x \rangle$ and $\langle y, z \rangle$ are primes minimal over I. Since $\langle x \rangle$ has codimension one and $\langle y, z \rangle$ has codimension two, the codimension of I is one.

The word *height* is sometimes used for codimension, but codimension suggests complementary dimension, which is geometrically sensible. For example, the geometric object $V(xy, xz)$ corresponds to a line $V(x)$ and a point $V(y, z)$ in $\mathbb{P}^2$; the line is defined by a single equation, so has dimension one less than the dimension of the ambient space. Notice that for the chain of primes $0 \subsetneq \langle x \rangle \subsetneq \langle x, y \rangle \subsetneq \langle x, y, z \rangle$ it is not clear that there could not be other, longer chains of primes.

Exercise 10.1.3. Codimension for some ideals in $k[x, y, z]$

1. Prove that the chain exhibited for $I = \langle x, y, z \rangle$ is maximal. Hint: localize and quotient.
2. What is the codimension of $\langle xy, xz, yz \rangle$?
3. How about of $\langle xy, x^2 \rangle$? $\diamond$

When commutative algebraists talk about the dimension of a ring R, they usually mean the *Krull dimension dim(R)*, which is the length of the longest chain of prime ideals contained in R. For example, a PID has Krull dimension one. It is obvious that a polynomial ring in n variables has dimension at least n (by induction). A polynomial ring over a field is *catenary*, which means that any two maximal chains of primes between prime ideals P and Q have the same length. So if R is a polynomial ring over a field, then

$$codim(I) = dim(R) - dim(R/I).$$

Our earlier remarks then show that $dim(k[x_0, \ldots, x_n]/I)$ is equal to the dimension of the affine variety $V(I) \subseteq \mathbb{A}^{n+1}$, so for homogeneous I the Krull dimension of R/I is one larger than the dimension of the projective variety $V(I) \subseteq \mathbb{P}^n$. For the example $I = \langle xy, xz \rangle \subseteq k[x, y, z]$, as an affine variety, $V(I)$ consists of a plane and a line, so has dimension two. As a projective variety, $V(I)$ is a line and a point, so has dimension one. For a proof that the Krull dimension of R modulo a homogeneous ideal is really one more than the dimension as defined in terms of the Hilbert polynomial, the best route is via Noether normalization, for example, see [3]. A nice exposition of the equivalence of the various definitions of dimension may be found in Balcerzyk and Jozefiak [6].

Definition 10.1.4. *The dimension of a finitely-generated R–module M is defined in terms of the annihilator ideal of M:*

$$dim(M) := dim(R/ann(M)).$$

As mentioned in Chapter 3, a graded ring R behaves very much like a local ring, where the ideal R_+ generated by all elements of positive degree plays the role of maximal ideal. So in the graded case, we'll call R_+ "the" maximal ideal m. Back in Chapter 3 we also defined a regular sequence: for a polynomial ring R and graded R-module M a regular sequence on M is just a sequence of homogeneous polynomials $\{f_1, \ldots, f_k\} \subseteq$ m such that f_1 is a nonzerodivisor on M and f_i is a nonzerodivisor on $M/\langle f_1, \ldots, f_{i-1} \rangle M$ for all $i > 1$. To define a regular sequence for a module over a local ring we simply drop the requirement of homogeneity (notice that in a local ring, any $f \notin$ m is a unit–see Exercise A.2.7). The *depth* of a module M over a graded or local ring is:

Definition 10.1.5. $depth(M) = sup\{j | \{f_1, \ldots, f_j\} \subseteq$ m *is a regular sequence on M*$\}$.

Example 10.1.6. Let $R = k[x_1, \ldots, x_n]$. If

$$M = R \oplus R/\mathrm{m},$$

then every nonconstant homogeneous element is a zero divisor, since R/m is a summand of M, so $depth(M) = 0$. Since R is a summand of M, the annihilator of M is zero, so $dim(M) = n$. Notice that the only prime ideal minimal over $ann(M)$ is the zero ideal, but $Ass(M) = \{(0), \mathrm{m}\}$. In particular, $Ass(M)$ can strictly contain the set of primes minimal over $ann(M)$.

Lemma 10.1.7. *If M is a finitely generated module over a Noetherian local ring R, then $depth(M) \leq dim(M)$.*

Proof. First, by Exercise 1.3.11 the union of the associated primes of M is the set of zero-divisors on M. Let $I = ann(M)$. The proof is by induction on the depth of M, with the base case obvious. Now suppose $\{f_1, \ldots, f_k\}$ is a maximal regular sequence on M. Then f_1 is a nonzerodivisor on M, so is not contained in any associated prime of M. In particular, f_1 is not contained in any prime minimal over I. Therefore the dimension of $R/\langle I, f_1 \rangle$ is strictly less than the dimension of $R/\langle I \rangle$: consider a maximal chain of primes

$$P_0 \subseteq P_1 \subseteq \cdots \subseteq P_n \subseteq R/\langle I \rangle.$$

The candidates for P_0 are just the minimal associated primes of I. Of course, a prime ideal in $R/\langle I, f_1 \rangle$ is just a prime ideal of $R/\langle I \rangle$ which contains f_1, so since f_1 is not contained in any minimal prime of I this means that

$$dim(R/\langle I, f_1 \rangle) < dim(R/\langle I \rangle).$$

But $depth(M/f_1 M) = k - 1$, which by the inductive hypothesis is at most $dim(R/\langle I, f_1 \rangle)$, so we're done. $\square$

Exercise 10.1.8. In the graded case, use a short exact sequence and Hilbert polynomials to give a quick proof of Lemma 10.1.7. $\diamond$

When $M = R$, it is common to write $depth(I)$ to mean the length of a maximal regular sequence (on R) contained in the ideal I; with this notation $depth(I) \leq codim(I)$. For $R = k[x_1, \ldots, x_n]$ we have $depth(R) = n$, and if $I \subseteq R$ then $depth(I) = codim(I)$. The main result relating depth and projective dimension is the famous Auslander–Buchsbaum theorem.

Theorem 10.1.9 (Auslander–Buchsbaum). *For a finitely generated module M over a Noetherian graded or local ring R, if $pdim(M)$ is finite, then*

$$pdim(M) + depth(M) = depth(R).$$

The proof requires some work; the interested reader can find complete details in [28]. On the other hand, it is fun to prove the following special case.

Exercise 10.1.10. Let M be a finitely generated graded module over $R = k[x_1, \ldots, x_n]$. Without using Auslander–Buchsbaum, prove that if

$\mathfrak{m} = \langle x_1, \ldots, x_n \rangle$ is an associated prime of M, then $depth(M) = 0$ and $pdim(M) = n$. Here is an outline: since $\mathfrak{m}$ is an associated prime of M, there exists $\alpha \in M$ such that $\mathfrak{m}$ is the annihilator of α. But α generates a submodule $R\alpha$ of M, so we have an exact sequence:

$$0 \longrightarrow R\alpha \longrightarrow M \longrightarrow M/R\alpha \longrightarrow 0.$$

Tensoring with $R/\mathfrak{m} \simeq k$ yields a long exact sequence of *Tor's*:

$$\cdots \longrightarrow Tor_{i+1}(k, M/R\alpha) \longrightarrow Tor_i(k, R\alpha) \longrightarrow Tor_i(k, M)$$
$$\longrightarrow Tor_i(k, M/R\alpha) \longrightarrow Tor_{i-1}(k, R\alpha) \longrightarrow \cdots$$

Now, we know that as an R-module $R\alpha \simeq k$, and the minimal free resolution of k over R is the Koszul complex, hence

$$Tor_n(k, R\alpha) \neq 0.$$

How does this complete the proof? ◇

10.2 Cohen–Macaulay Modules and Geometry

In this section, we study Cohen–Macaulay rings and modules; we begin with a remark of Mel Hochster (cited in [21]):

"Life is really worth living in a Cohen–Macaulay ring".

One reason is that there are nice connections to geometry; for example, all the local rings of a smooth variety are Cohen–Macaulay, and to combinatorics (see the next section).

Definition 10.2.1. *A finitely generated module M over a Noetherian local ring R is Cohen–Macaulay if depth$(M) = dim(M)$.*

A local ring R is Cohen–Macaulay if $depth(R) = dim(R)$, in other words, it is Cohen–Macaulay as a module over itself. A non-local ring is Cohen–Macaulay if localizing at any maximal ideal yields a (necessarily local) Cohen–Macaulay ring. In [53], a projective variety (or scheme) X is defined to be Cohen–Macaulay if all its local rings are Cohen–Macaulay; by [28], Proposition 18.8 this is true if the local rings at the points of X are Cohen–Macaulay. If $X \subseteq \mathbb{P}^n$ we say that X (or R/I_X) is *arithmetically Cohen–Macaulay* (abbreviated aCM) if the homogeneous coordinate ring R/I_X is a Cohen–Macaulay ring. This is true (see [28]) iff R/I_X is a Cohen–Macaulay $R = k[x_0, \ldots, x_n]$-module; it is this property that we now study. Notice that the Cohen–Macaulay property of a variety is *intrinsic* because it depends only on the local rings, whereas the aCM property is *extrinsic*, since it depends on how X sits in $\mathbb{P}^n$. We have that R/I_X is aCM iff

$$depth(R/I_X) = dim(R/I_X) = dim(R) - codim(I_X).$$

Since $depth(R) = n + 1 = dim(R)$, the Auslander–Buchsbaum theorem tells us that $pdim(R/I_X) = n + 1 - depth(R/I_X)$, so R/I_X is aCM iff

$$pdim(R/I_X) = codim(I_X).$$

Example 10.2.2. aCM examples

1. The twisted cubic C is aCM: in Section 3.2 we saw that $pdim(R/I_C) = 2$, and in Example 2.3.9 we found $HP(R/I_C, i) = 3i + 1$, so $codim(I_C) = 2$.

2.
```
i1 : R=ZZ/31991[x,y,z];

i2 : I=ideal(x*y,x*z,y*z);

o2 : Ideal of R

i3 : codim I

o3 = 2

i4 : pdim coker gens I

o4 = 2

i5 : res coker gens I

        1      3      2
o5 = R  <-- R  <-- R  <-- 0

        0      1      2      3

o5 : ChainComplex

i6 : o5.dd

           1                          3
o6 = 0 : R  <---------------- R   : 1
                | xy xz yz |
```

```
        3                               2
  1 :  R   <------------------  R   : 2
               {2} | -z  0  |
               {2} |  y  -y |
               {2} |  0   x |


        2
  2 :  R   <------ 0 : 3
            0
```

Notice that just like the twisted cubic, *I* is generated by the 2 by 2 minors of the syzygy matrix.

3. The ideal of six random points in $\mathbb{P}^2$ (From Chapter 7)

```
i1 : load "points"; R=ZZ/31991[x,y,z];
--loaded points

i3 : m=random(R^3,R^6)

o3 = | 9534   1166  405   156  9762  860 |
     | 7568   363   6195  484  2144  241 |
     | 14043  5756  1024  65   -563  -17 |

                3        6
o3 : Matrix R <--- R

i4 : res coker gens pointsideal1 m

        1      4      3
o4 = R   <-- R   <-- R   <-- 0

        0      1      2      3

o5 : betti o4

o5 = total: 1 4 3
          0: 1 . .
          1: . . .
          2: . 4 3

i6 : minors(3,o4.dd_2) == pointsideal1 m

o6 = true
```

Once again, the ideal is generated by the maximal minors of the matrix of first syzygies.

All three of these examples are evidence for the Hilbert-Burch theorem, which says if I is codimension 2 and R/I is aCM, then I is indeed generated by the maximal minors of the matrix of first syzygies. See [29], Chapter 3 for more.

A polynomial ring R over a field is Cohen–Macaulay [28]; we use this and the Auslander–Buchsbaum theorem to prove Theorem 8.3.2: If M is a finitely generated, graded R-module and P is a prime of codimension c, then $P \in Ass(M)$ iff $P \in Ass(Ext^c(M, R))$. First observe that if $P \in Ass(M)$, then depth $M_P = 0$. This follows since every non-unit in R_P is in P, and is thus a zero divisor on M.

Lemma 10.2.3. *In the setting above, if $P \in Ass(M)$ and $codim(P) = c$, then $Ext^i(M, R)_P = 0$ for $i > c$ and $Ext^c(M, R)_P \neq 0$.*

Proof. If $P \in Ass(M)$, then by Auslander–Buchsbaum and the observation above, $pdim(M_P) = depth(R_P)$; since R is Cohen–Macaulay, $depth(R_P) = c$. Thus, the free resolution of M_P over R_P goes back c steps, so obviously $Ext^c_{R_P}(M_P, R_P) \neq 0$, and $Ext^i_{R_P}(M_P, R_P) = 0$ if $i > c$. Localization commutes with Hom (exercise!), so we're done. $\square$

So, we still need to show that $P \in Ass(Ext^c(M, R))$. By Exercise 6.1.5, we know that P is in the support of $Ext^c(M, R)$, so P contains the annihilator of $Ext^c(M, R)$. If P properly contains a prime Q of codimension $< c$ minimal over the annihilator, then by Lemma 10.2.3, $Ext^c(M, R)_Q = 0$ which implies Q is not in the support of $Ext^c(M, R)$, hence not minimal over the annihilator, contradiction. Conclusion: P is a prime minimal over the annihilator of $Ext^c(M, R)$, so $P \in Ass(Ext^c(M, R))$.

Exercise 10.2.4. Finish the proof of Theorem 8.3.2 by showing that for a prime P of codimension c, $P \in Ass(Ext^c(M, R)) \Rightarrow P \in Ass(M)$. $\Diamond$

Exercise 10.2.5. Still assuming that M is a finitely generated, graded R-module, show that $i < codim(P)$ for all $P \in Ass(M) \Rightarrow Ext^i(M, R) = 0$. Combine this with Theorem 10.1.9 to show that M is Cohen–Macaulay iff $Ext^i(M, R) = 0$ for all but one value of i (can you see which?). $\Diamond$

One easy but important observation is that if R/I is aCM, then I cannot have embedded components; in fact, the associated primes of I must all have the same codimension. Let's take a look at an example with no embedded components, but for which $V(I)$ is not equidimensional. If $I = \langle xy, xz \rangle \subseteq k[x, y, z] = R$ we know the minimal primes of R/I are $\langle x \rangle$ and $\langle y, z \rangle$, so I has codimension 1. But it is obvious that the free resolution of R/I has projective dimension two, so R/I is not aCM.

Exercise 10.2.6. aCM or not?

1. $\langle x^2, xy \rangle \subseteq k[x, y]$
2. (Macaulay 2) In $\mathbb{Z}/101[x, y, z]$ the ideal $\langle x^2 - 19xy - 6y^2 - 35xz + 9yz - 44z^2, xy^2 - 13y^3 + xyz - 29y^2z - 25xz^2 + 15yz^2 - 50z^3, y^3 - 39xyz - 32y^2z - 23xz^2 - 20yz^2 + 44z^3 \rangle$
3. The image of the map $\mathbb{P}^1 \longrightarrow \mathbb{P}^4$ given by

$$(x : y) \longrightarrow (x^4 : x^3y : x^2y^2 : xy^3 : y^4). \quad \Diamond$$

It would be nice if equidimensional varieties $V(I)$ all had R/I aCM, but this is not true—here is an example of a smooth, *irreducible* projective variety which is not aCM:

Example 10.2.7. (The rational quartic) Map $\mathbb{P}^1 \overset{\phi}{\longrightarrow} \mathbb{P}^3$ via

$$(x : y) \longrightarrow (x^4 : x^3y : xy^3 : y^4).$$

Let $(a : b : c : d)$ be homogeneous coordinates for $\mathbb{P}^3$. The image X of $\phi(\mathbb{P}^1)$ is contained in the open sets U_a and U_d; on $U_a \cap X$ we define an inverse map via $(a : b)$, on $U_d \cap X$ we define an inverse map via $(c : d)$, so $\mathbb{P}^1$ and X are isomorphic. The codimension of I_X in $R = k[a, b, c, d]$ is therefore two. However, the projective dimension of R/I_X as an R-module is three. Since the Hilbert polynomial is $4t + 1$, Riemann-Roch provides (after checking smoothness, see Example A.3.2) another way to see that X is intrinsically a $\mathbb{P}^1$.

```
i1 : R=ZZ/101[a,b,c,d]; S=ZZ/101[x,y];

i3 : m=map(S,R,{x^4,x^3*y,x*y^3,y^4});

o3 : RingMap S <--- R
```

```
i4 : I=kernel m

                            3       2     2     2
o4 = ideal (b*c - a*d, c  - b*d , a*c  - b d,
                 3    2
                b  - a c)


o4 : Ideal of R

i5 : hilbertPolynomial coker gens I

o5 = - 3*P  + 4*P
            0       1


o5 : ProjectiveHilbertPolynomial

i6 : res coker gens I

          1       4       4       1
o6 = R  <-- R  <-- R  <-- R


        0       1       2       3


o6 : ChainComplex

i7 : o6.dd
                       1
o7 = -1 : 0 <------ R  : 0
                       0

              1                                              4
       0 : R  <-------------------------------- R  : 1
                    {0}|bc-ad b3-a2c ac2-b2d c3-bd2|


           4                              4
       1 : R  <-------------------- R  : 2
                    {2}|-b2 -ac -bd -c2|
                    {3}|c    d    0    0 |
                    {3}|a    b   -c   -d |
                    {3}|0    0    a    b |
```

```
        4               1
  2 : R   <------ R : 3
           {4}|d |
           {4}|-c|
           {4}|-b|
           {4}|a |
```

i8 : Ext^3(coker gens I, R)

o8 = cokernel {-5} | d c b a |

To understand this example, we need some definitions. Recall from Chapter 9 that a complete linear system on a curve is just the set of all global sections of the line bundle corresponding to a divisor, so is a finite dimensional vector space $H^0(\mathcal{O}(D))$. An incomplete linear system is a proper subspace of $H^0(\mathcal{O}(D))$. For example, on $\mathbb{P}^1$,

$$H^0(\mathcal{O}_{\mathbb{P}^1}(4)) \simeq span\{x^4, x^3y, x^2y^2, xy^3, y^4\}.$$

The homogeneous coordinate ring of the image of this map in $\mathbb{P}^4$ is isomorphic to $k[x^4, x^3y, x^2y^2, xy^3, y^4]$, while for the rational quartic, the homogeneous coordinate ring of the image in $\mathbb{P}^3$ is isomorphic to $k[x^4, x^3y, xy^3, y^4]$. How does this relate to the aCM property? The key is

Theorem 10.2.8 (Local duality, see [28]). *Let M be a finitely generated, graded $R = k[x_0, \ldots, x_n]$-module, and $i \geq 1$. Then as vector spaces,*

$$H^i(\widetilde{M}(m)) \simeq Ext^{n-i}(M, R)_{-m-n-1}.$$

In particular, for a curve $C \subseteq \mathbb{P}^n$ (even allowing several components, but all one-dimensional), $H^1(\mathcal{I}_C(m)) \simeq Ext^n(R/I_C, R)_{-m-n-1}$. Now, we know that a curve is codimension $n - 1$ in $\mathbb{P}^n$, so a curve will be aCM iff R/I_C has projective dimension $n - 1$ iff the only non-vanishing $Ext^i(R/I_C, R)$ occurs at $i = n - 1$, which means a curve in $\mathbb{P}^n$ is aCM iff for all m

$$Ext^n(R/I_C, R)_{-m-n-1} = H^1(\mathcal{I}_C(m)) = 0.$$

Why don't we have to worry about $Ext^{n+1}(R/I_C, R)$? Well, we can assume that I_C is saturated, so $m \notin Ass(R/I_C)$. Then by Exercise 10.1.10 *pdim* $(R/I_C) < n + 1$ and so $Ext^{n+1}(R/I_C, R) = 0$. The ideal sheaf sequence

$$0 \longrightarrow \mathcal{I}_C(m) \longrightarrow \mathcal{O}_{\mathbb{P}^n}(m) \longrightarrow \mathcal{O}_C(m) \longrightarrow 0$$

gives rise to

$$0 \longrightarrow H^0(\mathcal{I}_C(m)) \longrightarrow H^0(\mathcal{O}_{\mathbb{P}^n}(m)) \longrightarrow H^0(\mathcal{O}_C(m)) \longrightarrow H^1(\mathcal{I}_C(m))$$
$$\longrightarrow 0.$$

What this means is that C is aCM iff the map $H^0(\mathcal{O}_{\mathbb{P}^n}(m)) \longrightarrow H^0(\mathcal{O}_C(m))$ is a surjection iff R_m surjects onto $H^0(\mathcal{O}_C(m))$ for all m.

This example illustrates the wonderful fact that *Ext* modules can be used to compute sheaf cohomology; we return to the rational quartic X. From the free resolution we see that

$$Ext^3(R/I_X, R) \simeq R(5)/\langle a, b, c, d \rangle.$$

Thus $Ext^3(R/I_X, R)_i = 0$ unless $i = -5$, hence $H^1(\mathcal{I}_X(m))$ vanishes except at $m = 1$, and $H^1(\mathcal{I}_X(1))$ is a one dimensional vector space. We could also obtain this from the exact sequence in cohomology: $h^0(\mathcal{I}_X(1)) = 0$ since there is no linear relation on the sections used to embed X. By Riemann-Roch, $h^0(\mathcal{O}_X(1)) = 5$ and from Example 9.1.5, $h^0(\mathcal{O}_{\mathbb{P}^3}(1)) = 4$. As noted last chapter, a great description of the fine points may be found in Eisenbud's chapter in [95]; references for local duality are Bruns and Herzog [21], Brodmann and Sharp [19], and the appendix in [28].

Exercise 10.2.9. Prove that any hypersurface in $\mathbb{P}^n$ is aCM. Now find the flaw in the following argument. Map $\mathbb{P}^1 \xrightarrow{\phi} \mathbb{P}^2$ by a sublinear system of the linear system of Example 9.3.4 (in particular, we're using a divisor D of degree 3):

$$(x : y) \longrightarrow (x^3 : x^2y : y^3).$$

Let $(a : b : c)$ be homogeneous coordinates for $\mathbb{P}^2$; clearly the image Y of $\phi(\mathbb{P}^1)$ is simply $V(b^3 - a^2c)$. We have an exact sequence

$$0 \longrightarrow H^0(\mathcal{I}_Y(1)) \longrightarrow H^0(\mathcal{O}_{\mathbb{P}^2}(1)) \longrightarrow H^0(\mathcal{O}_Y(1)) \longrightarrow H^1(\mathcal{I}_Y(1)) \longrightarrow 0.$$

Now, there are no linear forms in I_Y, so $h^0(\mathcal{I}_Y(1)) = 0$, and $h^0(\mathcal{O}_{\mathbb{P}^2}(1)) = 3$. By Riemann-Roch, $h^0(\mathcal{O}_Y(1)) = h^0(\mathcal{O}_Y(D)) = deg(D) + 1 - g = 4$. So from the exact sequence, $h^1(\mathcal{I}_Y(1)) = 1$; hence Y is not aCM. But Y is a hypersurface, so what's wrong? $\diamond$

While we've got local duality on the table, we should revisit the concept of regularity. A coherent sheaf $\mathcal{F}$ on $\mathbb{P}^n$ is defined to be m-regular if

$$H^i(\mathcal{F}(m - i)) = 0$$

for all $i > 0$. In Chapter 7 we defined the regularity of a graded $R = k[x_0, \ldots, x_n]$-module N in terms of a minimal free resolution of N; in Macaulay 2 terms the regularity of N corresponds to the label of the bottom row of the betti diagram of N. For a coherent sheaf $\mathcal{F}$ as above, we obtain a graded R-module by taking the direct sum of all the global sections $\Gamma_*(\mathcal{F}) = \oplus_i H^0(\mathcal{F}(i))$. How does the regularity of $\Gamma_*(\mathcal{F})$ relate to the regularity of $\mathcal{F}$? Alas, if $\Gamma_*(\mathcal{F})$ has zero-dimensional associated primes, then it cannot be finitely generated as an R-module:

Example 10.2.10. Let $X = (0 : 0 : 1) \subseteq \mathbb{P}^2$, and let $\mathcal{F} = \mathcal{O}_X$. There are several ways to see that $h^0(\mathcal{O}_X(m)) = 1$ for all m negative (think about why this means $\Gamma_*(\mathcal{O}_X)$ cannot be finitely generated). First we have the intrinsic description: since X is a point, regular functions are just constants, and the twist is irrelevant. Alternately, since $h^0(\mathcal{I}_X(m)) = h^0(\mathcal{O}_{\mathbb{P}^2}(m)) = 0$ for m negative, we have that $h^0(\mathcal{O}_X(m)) = h^1(\mathcal{I}_X(m))$, and by local duality this last number is the dimension of $Ext^2(R/\langle x, y \rangle, R)_{-m-3} \simeq k[z]_{-m-1}$, which is one dimensional for all $m \leq -1$. A third alternative is to observe that the nonvanishing of $H^1(\mathcal{I}_X(m))$ for m negative follows from the results of Chapter 7.

This is very disconcerting – after all, a word should really have a single meaning! Happily, it turns out ([29], Chapter 4) that as long as $\Gamma_*(\mathcal{F})$ does not have zero-dimensional associated primes the two definitions of regularity actually do agree. We illustrate this for a set of points $X \subseteq \mathbb{P}^n$. It is obvious that I_X has (as a module) no associated primes. In Chapter 7 we asserted that the regularity as defined via the betti diagram of I_X coincided with the smallest i such that $H^1(I_X(i-1)) = 0$. If I_X is a saturated ideal of codimension n, then obviously $Ext^i(R/I_X, R) \neq 0$ iff $i = n$, thus R/I_X is aCM. If j is the label of the bottom row of the betti diagram for R/I_X, then $Ext^n(R/I_X, R)$ has a generator of degree $-n - j$. When X is zero-dimensional, local duality says that $Ext^n(R/I_X, R)_{-i-n} \simeq H^1(\mathcal{I}_X(i-1))$, so I_X will be m-regular when $Ext^n(R/I_X, R)_{-j-n} = 0$ for all $j \leq m$. For the example of twenty generic points in $\mathbb{P}^3$ we computed the free resolution:

$$0 \longrightarrow R^{10}(-6) \longrightarrow R^{24}(-5) \longrightarrow R^{15}(-4) \longrightarrow R \longrightarrow R/I_X \longrightarrow 0.$$

Since $Ext^3(R/I_X, R)$ is generated in degree -6, $Ext^3(R/I_X, R)_{-7} = 0$, and we see that I_X is four regular, just as expected.

To recap, the basic idea is that the regularity as defined in 7.1.7 corresponds to the largest j such that $Ext^i(M, R)_{-i-j}$ is nonzero (taken over all i). By local duality, this corresponds to the nonvanishing of a certain sheaf cohomology module, which is how we defined the regularity of a set of points. We end this section with a famous conjecture on regularity. The conjecture is known to be true in certain cases [50], [60], but is open in general.

Conjecture 10.2.11 (Eisenbud-Goto, [30]). *Let* $P \subset k[x_0, \ldots, x_n]$ *be a homogeneous prime ideal containing no linear form. Then*

$$reg(P) \leq degree(R/P) - codim(P) + 1.$$

10.3 The Upper Bound Conjecture for Spheres

We return to a question raised in Chapter 5: If we fix the number of vertices, what is the biggest possible f-vector for a d-dimensional simplicial polytope? We compare f-vectors pointwise, so $f(\Delta_1) \geq f(\Delta_2)$ if

$$f_i(\Delta_1) \geq f_i(\Delta_2)$$

for all i. More generally, consider a simplicial $(d-1)$-sphere (a triangulation of S^{d-1}): what is the biggest possible f-vector for such a triangulation? Rather surprisingly, there are simplicial $(d-1)$-spheres which *do not* correspond to simplicial polytopes. By the Dehn–Sommerville relations, if we know $f_0, \ldots, f_{\lfloor \frac{d}{2} \rfloor}$, then we know the entire f-vector. The convex hull of n distinct points on the moment curve

$$(t, t^2, t^3, \ldots, t^d)$$

turns out to be a simplicial d-polytope with n vertices; it is called a cyclic polytope and denoted $C_d(n)$. It is good fun (see Chapter 0 of Ziegler) to show that

$$f_i(C_d(n)) = \binom{n}{i+1}, \quad i < \left\lfloor \frac{d}{2} \right\rfloor.$$

Conjecture 10.3.1 (Motzkin). *Let P be a triangulation of S^{d-1} having n vertices. Then for all i,*

$$f_i(P) \leq f_i(C_d(n)).$$

At first this seems trivial, because for a fixed n and d, $C_d(n)$ obviously maximizes the first "half" of the f-vector. The content of Motzkin's conjecture is that maximizing the first half of the f-vector also maximizes the remaining half. McMullen gave a slick combinatorial proof for the special case of simplicial polytopes in 1970 using a technique called shelling ([100], Chapter 8). In 1974, Stanley gave a beautiful algebraic proof of the conjecture, which we now sketch. For a full treatment, see either [88] or [89]. The first step is to reformulate the conjecture in terms of the h-vector (which we encountered in Chapter 5); recall that the Dehn–Sommerville relations say that the h-vector of a simplicial polytope is symmetric.

Conjecture 10.3.2 (Reformulation of 10.3.1). *Let P be a triangulation of S^{d-1} having n vertices. Then*

$$h_i(P) \leq \binom{n - d + i - 1}{i}.$$

A key ingredient in Stanley's proof is Reisner's theorem. To state Reisner's theorem we need to define the link of a face $F \subseteq \Delta$:

$$lk(F) = \{G \in \Delta \,|\, G \cap F = \emptyset, G \cup F \in \Delta\}.$$

Theorem 10.3.3 (Reisner). *If Δ is a simplicial complex with n vertices, then the Stanley–Reisner ring $k[v_1, \ldots, v_n]/I_\Delta$ is aCM iff for each face F and $i < \dim lk(F)$,*

$$\widetilde{H}_i(lk(F)) = 0,$$

where the homology is computed with coefficients in k.

Notice that *the field matters*. For the proof of Theorem 10.3.3 we refer to Reisner's original paper [79] (which is a great advertisement for Frobenius and characteristic p methods); an alternate treatment appears in [21]. If Δ is a triangulation of a manifold, then for any nonempty face F, $lk(F)$ is either a homology sphere or a homology cell–the adjective homology here means that its nonreduced homology looks like that of a sphere (nonzero only at top and bottom) or of a cell (nonzero only at bottom). In particular *the Stanley–Reisner ring of a simplicial sphere is aCM*. Note that the primary decomposition of the Stanley-Reisner ring which appeared in Theorem 5.3.3 provides some evidence for this: for a simplicial sphere, clearly all the minimal cofaces have the same number of vertices, so that $V(I_\Delta)$ is equidimensional, which is consistent with the remarks in Section 10.1. To prove the theorem, we need the following characterization of the aCM property.

Lemma 10.3.4 (Hironaka's criterion). *Let $R = k[x_1, \ldots, x_n]$, k an infinite field, I a homogeneous ideal, and suppose R/I has Krull dimension d. Then R/I is aCM iff there exists a regular sequence of linear forms $l_1, \ldots, l_d$ and* *homogeneous elements* $v_1, \ldots, v_m$ *such that any* $f \in R/I$ *may be written uniquely as*

$$f = \sum_{i=1}^{m} v_i \, p_i(l_1, \ldots, l_d).$$

Corollary 10.3.5 (Stanley). *If* R/I *is aCM with* v_i *as above, then the Hilbert series may be written:*

$$P(R/I, t) = \frac{\sum_{i=1}^{m} t^{\text{degree } v_i}}{(1 - t)^d}.$$

Exercise 10.3.6. Prove the corollary, given the lemma. ◇

We know that chopping down by a regular sequence will preserve the numerator of the Hilbert series, so the Hilbert series of $R' = R/\langle I, l_1, \ldots, l_d \rangle$ is the numerator of the Hilbert series of R/I. In Chapter 5 we saw that if $I = I_\Delta$, then this numerator is $h(\Delta)$. In the situation we're interested in, I_Δ will have no degree one elements, so h_1 of R' is $n - d$, and after a change of variables we can think of R' as the quotient of a polynomial ring in $n - d$ variables. Since a polynomial ring S in $n - d$ variables has $HF(S, i) = \binom{n-d+i-1}{i}$ and R' is a quotient of such a ring, the h_i must satisfy

$$h_i \leq \binom{n - d + i - 1}{i}.$$

This proves the upper bound conjecture for simplicial spheres! Can you see what happens if equality holds?

Exercise 10.3.7. Write a program (without using your code from Exercise 5.3.5) which takes as input a simplicial complex and returns the h-vector. Hint: find a way to express the relationship between the h-vector and f-vector in terms of univariate polynomials. ◇

For the combinatorially inclined reader, it is worth mentioning that there are all sorts of open questions on the f-vectors of polytopes; for example, the f-vectors of non-simplicial four polytopes are not classified. For simplicial polytopes, there is a complete characterization of possible f-vectors, due to Billera–Lee [16] and Stanley [90] (using toric methods). One consequence of this is my favorite counterexample in mathematics (see [17] or [61]), which we close with:

Conjecture 10.3.8 (Motzkin). *The f-vector of a simplicial polytope is unimodal.*

The conjecture is true in all dimensions ≤ 19, but in dimension twenty there is a simplicial polytope with

$$4, 203, 045, 807, 626$$

vertices whose f-vector is not unimodal! So your intuition in low dimension can lead you astray.

Supplemental Reading: As usual, Eisenbud [28] is excellent; there are nice short treatments in Matsumura [64] and Sharp [84]. The most comprehensive source is the book [21] by Bruns and Herzog. My use of local duality in Section 10.2 is a bit unfair–I intended to write an appendix on this, but everything I'd say already appears in Chapter 9 of [29]. In fact, local duality is usually stated in terms of *local cohomology*. Here's the gist: if R is Noetherian, I an ideal and M an R-module, then we set

$$H_I^0(M) = \{m \in M \mid I^d m = 0, \text{ some } d\}.$$

H_I^0 is a left exact, covariant functor; $R^i H_I^0(M)$ is the ith local cohomology of M. For the specific case when M is a finitely generated, graded $R = k[x_0, \ldots, x_n]$-module and $\mathrm{m} = R_+$, *local duality* [29] tells us that

$$HF(Ext^{n+1-i}(M, R), -j - n - 1) = HF(H_{\mathrm{m}}^i(M), j).$$

(again, I fudge here a bit, if you're careful there are dual vector spaces involved, but for dimension counts that's not important). We check this on the rational quartic X, where we computed $Ext^3(R/I_X, R)_{-5}$; from the long exact sequence in Ext this is the same thing as $Ext^2(I_X, R)_{-5}$. In Macaulay 2, local cohomology is obtained with HH:

```
i10 : hilbertFunction(-5,Ext^2(image gens I, R))
```

```
o10 = 1
```

```
i11 : hilbertFunction(1,HH^2(image gens I))
```

```
o11 = 1
```

An argument with the Cech complex (see section A.4.1 of [28]) shows that if $i > 1$ then

$$HF(H_\mathfrak{m}^i(M), j) = dim_k H^{i-1}(\widetilde{M}(j)),$$

which gives us the version of local duality in Theorem 10.2.8. Back to supplemental reading: for the last section, see the original papers of Stanley and Reisner; Chapter 5 of Bruns and Herzog, Chapter 3 of Stanley [88] and Chapters 1 and 8 of Ziegler [100].

Appendix A

Abstract Algebra Primer

A.1 Groups

Let G be a set of elements endowed with a binary operation $G \times G \longrightarrow G$. We will write $\cdot$ for this operation. By definition, G is closed under the operation. G is a group if:

1. $\cdot$ is associative: $(a \cdot b) \cdot c = a \cdot (b \cdot c)$.
2. G possesses an identity element e such that $\forall\, g \in G, g \cdot e = e \cdot g = g$.
3. Every $g \in G$ has an inverse g^{-1} such that $g \cdot g^{-1} = g^{-1} \cdot g = e$.

The operation $\cdot$ need not be commutative; if it is then G is *abelian*. For abelian groups, the group operation is often written as $+$. For example,

$$\mathbb{Z}/n\mathbb{Z}$$

is an abelian group with the usual operation of addition modulo n. Notice if we define the operation $+$ as multiplication modulo n, then in general $\mathbb{Z}/n\mathbb{Z}$ will not be a group: for example, in $\mathbb{Z}/6\mathbb{Z}$ the class of 2 has no inverse. Are there values of n for which $\mathbb{Z}/n\mathbb{Z} - \{0\}$ is a group under multiplication?

Everyone's first example of a nonabelian group is the set of permutations of the numbers $\{1, 2, 3\}$. If we label the vertices of an equilateral triangle as $\{1, 2, 3\}$ then we may also think of this as the set of rigid motions of the triangle. The operations are rotation by $\frac{2\pi}{3}$ and reflection about a line connecting a vertex to the midpoint on the opposite edge. Think of a tuple (i, j, k) as representing the permutation $i \to j, j \to k, k \to i$; for example $(1, 2, 3)$ represents the permutation $1 \to 2, 2 \to 3, 3 \to 1$, whereas $(2, 3)$ represents $2 \to 3, 3 \to 2$, and 1 is left fixed. The group operation is just composition $\circ$. With our notation we can compose two permutations by tracing their composed action. For example, to compute

$$(1, 2, 3) \circ (2, 3),$$

163

we start reading from the right: $2 \to 3$ now step left, where $3 \to 1$. So the effect of the composition is to send $2 \to 1$. Now that we know where 2 goes, let's see where 3 goes: again starting from the right, $3 \to 2$, step left, where $2 \to 3$. So 3 is fixed by the composition. Finally, $(2, 3)$ fixes 1, step left, where 1 is sent to 2, so the composition sends $1 \to 2$. We conclude

$$(1, 2, 3) \circ (2, 3) = (1, 2).$$

Exercise A.1.1. The group of permutations on $\{1, 2, 3\}$ is denoted S_3. Write out the table for the group law. Then interpret it geometrically in terms of the rigid motions of the triangle. ◇

A *subgroup* of a group is just a subset which is itself a group; a particularly important class of subgroups are the normal subgroups. A subgroup H of G is normal if $gHg^{-1} \subseteq H$ for all $g \in G$, this condition means that the quotient G/H is itself a group. A *homomorphism* of groups is a map which preserves the group structure, so a map $G_1 \xrightarrow{f} G_2$ is a homomorphism if

$$f(g_1 \cdot g_2) = f(g_1) \cdot f(g_2).$$

The *kernel* of f is the set of $g \in G_1$ such that $f(g) = e$.

Exercise A.1.2. Prove that the kernel of a homomorphism is a normal subgroup. ◇

If at this point you are feeling confused, then you should go back and review some basic algebra. If you are yawning and saying "ho hum, seen all this stuff", that is good, because now we come to another topic covered in a first abstract algebra class, but which is not usually emphasized.

A.2 Rings and Modules

A *ring* is an abelian group under addition ($+$), with an additional associative operation multiplication ($\cdot$) which is distributive with respect to addition. Think for a moment and find examples of a noncommutative ring, and a ring without unit. All the rings we consider will also have a multiplicative identity (denoted by 1), and the multiplication will be commutative. If every element in a ring (save the additive identity 0) has a multiplicative inverse, then the ring is a *field*. A (nonzero) element a is called a zero divisor if there is a (nonzero) element b with $a \cdot b = 0$; a ring with no zero divisors is called a *domain* or integral domain. For emphasis, we repeat: in this book, *ring* means *commutative ring with unit*.

Example A.2.1. Examples of rings.

1. $\mathbb{Z}$, the integers.
2. $\mathbb{Z}/n\mathbb{Z}$, the integers mod n.
3. $A[x_1, \ldots, x_n]$, the polynomials with coefficients in a ring A.
4. $C^0(\mathbb{R})$, the continuous functions on $\mathbb{R}$.
5. k a field.

In linear algebra, we can add two vectors together, or multiply a vector by an element of the field over which the vector space is defined. Module is to ring what vector space is to field. Formally, a *module M* over a ring R is an abelian group, together with action of R on M which is R-linear: for $r_i \in R$, $m_i \in M$, $r_1(m_1 + m_2) = r_1 m_1 + r_1 m_2$, $(r_1 + r_2)m_1 = r_1 m_1 + r_2 m_1$, $r_1(r_2 m_1) = (r_1 r_2)m_1$, $1(m) = m$. An R-module M is *finitely-generated* if $\exists \{m_1, \ldots, m_n\} \subseteq M$ such that any $m \in M$ can be written $m = \sum_{i=1}^{n} r_i m_i$ for some $r_i \in R$. The most important class of modules are *ideals*: submodules of the ring itself.

Exercise A.2.2. Which are ideals?

1. $\{f \in C^0(\mathbb{R}) \mid f(0) = 0\}$.
2. $\{f \in C^0(\mathbb{R}) \mid f(0) \neq 0\}$.
3. $\{n \in \mathbb{Z} \mid n = 0 \bmod 2\}$.
4. $\{n \in \mathbb{Z} \mid n \neq 0 \bmod 2\}$. ◇

Example A.2.3. Examples of modules over a ring R.

1. Any ring is a module over itself.
2. A quotient ring R/I is both an R-module and an R/I-module.
3. An R-module M is *free* if M is isomorphic to a direct sum of copies of R; in particular M is free of relations. To see that this is not the case in general, suppose we have a ring R, and consider a free module M consisting of two copies of R, with generators ϵ_1 and ϵ_2. Elements of M may thus be written as two by one vectors with entries in R, with

module operations performed just as we do them in linear algebra. Now let's add a twist. Pick a nonzero $m \in M$, which we can write as

$$\begin{bmatrix} r_1 \\ r_2 \end{bmatrix}$$

for some $r_i \in R$. m generates a (principal) submodule $\langle R \cdot m \rangle \subseteq M$, so we can form the quotient module

$$M' = M/\langle R \cdot m \rangle.$$

This is obviously not a free module, since $\epsilon_1 = \binom{1}{0}$ and $\epsilon_2 = \binom{0}{1}$ are nonzero elements of M', but there is a relation between them: in M', $r_1 \cdot \epsilon_1 + r_2 \cdot \epsilon_2 = 0$.

If A and B are rings, a ring homomorphism $\phi : A \rightarrow B$ is a map such that:

$$\phi(a \cdot b) = \phi(a) \cdot \phi(b)$$
$$\phi(a + b) = \phi(a) + \phi(b)$$
$$\phi(1) = 1.$$

Let M_1 and M_2 be modules over A, $m_i \in M_i$, $a \in A$. A homomorphism (or map) of A-modules $\psi : M_1 \rightarrow M_2$ is a function ψ such that

$$\psi(m_1 + m_2) = \psi(m_1) + \psi(m_2)$$
$$\psi(a \cdot m_1) = a \cdot \psi(m_1).$$

An important instance of both ring and module maps is the map from a ring A to a quotient A/I. Is the kernel of a ring map (of rings with unit) a ring (with unit)? Given a homomorphism of A-modules $\psi : M_1 \rightarrow M_2$, the kernel consists of those elements of M_1 sent to zero by ψ. The image of ψ is the set $\{m \in M_2 | m = \psi(n)\}$ (the elements of M_2 "hit" by ψ) and the cokernel of ψ is $M_2/\psi(M_1)$. It is easy to check (do so if this is unfamiliar!) that the kernel, image, and cokernel are all A-modules.

Exercise A.2.4. Let $R = k[x, y, z]$ and define a map of modules $R^3 \xrightarrow{\phi} R^1$, where ϕ is the three by one matrix $[x, y, z]$. The kernel of ϕ is generated by the columns of the matrix ψ:

$$\begin{bmatrix} y & z & 0 \\ -x & 0 & z \\ 0 & -x & -y \end{bmatrix}$$

This is just linear algebra, but with matrices of polynomials. Prove that the kernel of ψ is not a free module, i.e. find a (polynomial) relation between the columns of ψ. Hint: just write down a polynomial vector

$$\begin{bmatrix} f_1 \\ f_2 \\ f_3 \end{bmatrix},$$

multiply it against ψ, and see what relations the f_i must satisfy. Check your solution against the solution (in Macaulay 2 syntax) given in the next section (but take ten minutes to try it yourself first!) ◇

Exercise A.2.5. Types of ideals and geometry

1. An ideal I is *principal* if I can be generated by a single element.
2. An ideal $I \neq \langle 1 \rangle$ is *prime* if $f \cdot g \in I$ implies either f or g is in I.
3. An ideal $I \neq \langle 1 \rangle$ is *maximal* if there does not exist any proper ideal J with $I \subsetneq J$.
4. An ideal $I \neq \langle 1 \rangle$ is *primary* if $f \cdot g \in I$ implies either f or g^m is in I, for some $m \in \mathbb{N}$.
5. An ideal I is *irreducible* if there do not exist ideals J_1, J_2 such that $I = J_1 \cap J_2, I \subsetneq J_i$.
6. An ideal I is *radical* if $f^m \in I$ ($m \in \mathbb{N}$) implies $f \in I$.

In $R = k[x, y]$, which classes (above) do the following ideals belong to? It may be helpful to read a bit of Chapter 1 first, and then draw a picture of the corresponding variety before tackling these. After you have read the section of Chapter 1 on primary decomposition, see what the Macaulay 2 command `primaryDecomposition` tells you.

1. $\langle xy \rangle$
2. $\langle y - x^2, y - 1 \rangle$
3. $\langle y, x^2 - 1, x^5 - 1 \rangle$
4. $\langle y - x^2, y^2 - yx^2 + xy - x^3 \rangle$
5. $\langle xy, x^2 \rangle$ ◇

Exercise A.2.6. Prove that a maximal ideal is prime. ◇

Exercise A.2.7. A local ring is a ring with a unique maximal ideal m. Prove that in a local ring, if $f \notin$ m, then f is a unit. ◇

Exercise A.2.8. If I is an ideal, when is R/I a domain? A field? A ring is called a principal ideal domain (PID) if it is an integral domain, and every ideal is principal. Go back to your abstract algebra textbook and review the

Euclidean algorithm, and use it to show that $k[x]$ is a PID. Find the generator of the ideal $\langle x^4 - 1, x^3 - 3x^2 + 3x - 1 \rangle$. Is $k[x, y]$ a PID? $\Diamond$

We close this section with the definition of a *direct limit*. A directed set S is a partially ordered set with the property that if $i, j \in S$ then there exists $k \in S$ with $i \leq k, j \leq k$. Let R be a ring and $\{M_i\}$ a collection of R-modules, indexed by a directed set S, such that for each pair $i \leq j$ there exists a homomorphism $\mu_{ji} : M_i \to M_j$. If $\mu_{ii} = id_{M_i}$ for all i and $\mu_{kj} \circ \mu_{ji} = \mu_{ki}$ for all $i \leq j \leq k$, then the modules M_i are said to form a directed system. Given a directed system, we build an R-module (the direct limit) as follows: let N be the submodule of $\oplus M_l$ generated by the relations $m_i - \mu_{ji}(m_i)$, for $m_i \in M_i$ and $i \leq j$. Then the direct limit is

$$\lim_{\to} M_l = \oplus M_l / N.$$

This is a pretty simple concept: we identify elements $m_i \in M_i$ and $m_j \in M_j$ if the images of m_i and m_j eventually agree.

A.3 Computational Algebra

Macaulay 2 is a computer algebra system, available at

```
http://www.math.uiuc.edu/Macaulay2.
```

It is available for basically all platforms; some tips for installing it (and general commutative algebra news) can be found at:

```
http://www.commalg.org/
```

To start it on a unix machine, just type M2 on the command line (make sure your paths are set). Macaulay 2 will respond with

```
Macaulay 2, version 0.9.2
--Copyright 1993-2001, D. R. Grayson and
  M. E. Stillman
--Singular-Factory 1.3b, copyright 1993-2001,
  G.-M. Greuel, et al.
--Singular-Libfac 0.3.2, copyright 1996-2001,
  M. Messollen

i1 :
```

Let's do Exercise A.2.4. First we make a polynomial ring – we'll use $\mathbb{Z}/101$ as the base field, but you can also work over other finite fields, the rationals, and more. Input lines are prefixed with i, and output lines with o; input is ended with a return. Try the following:

```
i1 : R=ZZ/101[x,y,z]

o1 = R

o1 : PolynomialRing

i2 : M=matrix {{x,y,z}}

o2 = | x y z |

                1        3
o2 : Matrix R   <--- R

i3 : kernel M

o3 = image {1} | 0   -y -z |
           {1} | -z  x   0  |
           {1} | y   0   x  |

                                  3
o3 : R-module, submodule of R
```

We make a matrix with a list of lists; once we have a matrix we can ask for the kernel, which we did on line i3. Of course, the kernel of *M* is a submodule of R^3; we want to convert the submodule on line o3 into a matrix, we do this by asking for the generators of the module:

```
i4 : gens o3

o4 = {1} | 0   -y -z |
     {1} | -z  x   0  |
     {1} | y   0   x  |

            3       3
o4 : Matrix R   <--- R
```

```
i5 : kernel o4

o5 = image {2} | x  |
           {2} | z  |
           {2} | -y |

                                  3
o5 : R-module, submodule of R
```

So we have solved our problem. Of course, there are many other ways we could have solved the problem, for example we could also have typed:

```
i6 : kernel matrix {{0,-y,-z},{-z,x,0},{y,0,x}}

o6 = image {1} | x  |
           {1} | z  |
           {1} | -y |

                                  3
o6: R-module, submodule of R
```

The numbers at the left of the matrix are related to *grading*, which is discussed in Chapter 2. The numbers differ because the matrix on line o4 maps $R^3(-2)$ to $R^3(-1)$,

```
i7 : degrees source o4

o7 = {{2}, {2}, {2}}

i8 : degrees target o4

o8 = {{1}, {1}, {1}}
```

reflecting the fact that the matrix was obtained as the kernel of the matrix o2. If we hand Macaulay 2 a matrix with no specifications, the target is assumed to have degree zero generators; so the matrix on line o6 maps $R^3(-1)$ to R^3.

Example A.3.1. There are three main loop structures in Macaulay 2: they are scan, apply, and while-do; all are documented online.

1. The apply command expects as input a list l and a function f, it returns the list $f(l)$:

```
i1 : apply({2,3,5,6}, i->i^2)

o1 = {4, 9, 25, 36}
```

2. The scan command is similar to apply, except that when f is applied to l, the result is not saved anywhere unless specified by the user.

```
i2 : scan({2,3,5,6}, i->i^2)

i3 : scan({2,3,5,6}, i-><<" "<< i^2)
4 9 25 36
```

(the << are a way of printing data all on the same line).
3. The while loop

```
i1 : i = 0;

i2 : while i < 10 do (<< " " << i; i = i + 1)
0 1 2 3 4 5 6 7 8 9
```

Example A.3.2. Here is a more substantive example. Suppose we are asked to study (whatever that means!) the image of a random degree d map from $\mathbb{P}^1$ to $\mathbb{P}^n$ (a *random rational curve*). If $C \subseteq \mathbb{P}^n$ is an irreducible curve (so $I_C = \langle f_1, \ldots, f_k \rangle$ is a prime ideal), then a point $p \in C$ is *singular* if the rank of the Jacobian matrix (evaluated at p) is less than $n - 1$ (which is the codimension of C). In other words, the singular locus is defined by I_C and the $n - 1 \times n - 1$ minors of

$$
\begin{bmatrix}
\frac{\partial f_1}{\partial x_0} & \frac{\partial f_1}{\partial x_1} & \cdots & \frac{\partial f_1}{\partial x_n} \\
\frac{\partial f_2}{\partial x_0} & \frac{\partial f_2}{\partial x_1} & \cdots & \frac{\partial f_2}{\partial x_n} \\
\vdots & \vdots & \vdots & \vdots \\
\frac{\partial f_k}{\partial x_0} & \frac{\partial f_k}{\partial x_1} & \cdots & \frac{\partial f_k}{\partial x_n}
\end{bmatrix}.
$$

A point $p \in C$ is called *smooth* if the rank of the Jacobian matrix at p is $n - 1$; C is smooth if every point of C is a smooth point. When is a random rational curve of degree d in $\mathbb{P}^n$ smooth? If it is not smooth, what is the singular locus?

```
issmooth = (I)->(c = codim I;
                 J = jacobian mingens I;
                 minors(c,J)+I)
```

```
--return the ideal defining the singular locus of
--a curve
```

```
randratcurve =  (d,n)->(R=ZZ/31991[s,t];
                        rands = random(R^{d}, R^(n+1));
                        --get n+1 elements of R_d
                        S=ZZ/31991[a_0..a_n];
                        kernel map(R,S,rands))
```

```
--given a degree d and target space P^n, find the
--ideal of a random rational curve. If d < n, then
--the image will lie in a linear subspace of P^n.
```

```
iexamples = (i,d,n)->(
        apply(i, j->(
           slocus = issmooth(randratcurve(d,n));
        degree coker gens slocus)))
--run i examples.
```

When you run this code, you'll see that as long as $n \geq 3$, the image of the curve is smooth. What follows is argued in detail in Hartshorne section IV.3, so we only give a quick sketch. First, picking a random rational curve of degree d in $\mathbb{P}^n$ corresponds to picking a generic $(n + 1)$-dimensional subspace of $H^0(\mathcal{O}_{\mathbb{P}^1}(d))$, which in turn corresponds to a generic projection of C from $\mathbb{P}^d$ to $\mathbb{P}^n$. For any curve C in $\mathbb{P}^n$, $n > 3$ the secant variety should have dimension three: there are two degrees of freedom to pick points p_1, p_2 on C, and then another degree of freedom in choosing a point on the line $\overline{p_1 p_2}$. So if $n > 3$, the secant variety of the curve will not fill up $\mathbb{P}^n$. This means that a generic point q will not lie on the locus of secant lines, so that projection from q will be one-to-one (exercise: show the locus of tangent lines has dimension at most two). To see that the image is actually smooth requires just a bit more effort, and we refer to [53], Proposition IV.3.4. Thus, any curve can be embedded in $\mathbb{P}^3$. What if we project all the way down to $\mathbb{P}^2$?

Save the code on the previous page in a file (e.g., A32)

```
i3 : load "A32";
--loaded A32

i4 : iexamples(5,3,2)

o4 = {1, 1, 1, 1, 1}

i5 : iexamples(5,4,2)

o5 = {3, 3, 3, 3, 3}

i6 : iexamples(5,5,2)

o6 = {6, 6, 6, 6, 6}

i7 : iexamples(5,6,2)

o7 = {10, 10, 10, 10, 10}
```

So a random projection of a degree d rational curve to $\mathbb{P}^2$ results in a curve with $\binom{d-1}{2}$ singular points. A singular point of a plane curve is called a *node* if it consists of two irreducible smooth branches crossing transversely. For example, the curve $V(y^2z - x^3 - x^2z) \subseteq \mathbb{P}^2$ has a node at the point $(0 : 0 : 1)$; if we plot the curve on the affine patch where $z = 1$ the picture is:

For plane curves with only nodal singularities, there is a *genus formula*:

Theorem A.3.3. *If $C \subseteq \mathbb{P}^2$ is an irreducible curve of degree d with only (δ) nodes as singularities, then the genus of the desingularization $\widetilde{C}$ is given by*

$$g(\widetilde{C}) = \binom{d-1}{2} - \delta.$$

For a random degree d planar rational curve, our computations show that there are $\binom{d-1}{2}$ singular points, which agrees exactly with the genus formula (of course, a computation is not a proof, and it remains to check that the singular points are really nodes). The genus formula is itself a consequence of the Riemann-Hurwitz formula; nice explanations of both can be found in Griffiths [48]. Finally, it should be noted that there is a local description of smoothness:

Definition A.3.4. *Let R be a Noetherian local ring with maximal ideal* m. *R is a regular local ring if*

$$\dim_{R/\mathfrak{m}} \mathfrak{m}/\mathfrak{m}^2 = \dim R.$$

Theorem A.3.5. *$p \in C$ is a smooth point iff the local ring at p is a regular local ring.*

See [53], section 1.5 for the proof.

Exercise A.3.6. Prove that the ring $k[x, y, z]/\langle y^2z - x^3 - x^2z \rangle$, localized at the ideal $\langle x, y \rangle$, is not a regular local ring. What point of the curve does the ideal $\langle x, y \rangle$ correspond to? ◇

There are several other computer algebra packages which have similar functionality to Macaulay 2; Singular is available at

http://www.singular.uni-kl.de/

Greuel and Pfister have written a commutative algebra book [47] based on Singular. CoCoA is another system, available at

http://cocoa.dima.unige.it/

Kreuzer and Robbiano have written a commutative algebra book [58] based on CoCoA. Supplemental Reading: some options for additional algebra background are Hungerford [56], Lang [59] or Rotman [81].

Appendix B
Complex Analysis Primer

B.1 Complex Functions, Cauchy–Riemann Equations

Let x and y be real numbers. A complex number is a number of the form

$$z = x + iy,$$

where $i^2 = -1$. The complex numbers form a field $\mathbb{C}$, which is the *algebraic closure* of the real numbers; in other words every non-constant univariate polynomial with coefficients in $\mathbb{C}$ has a root in $\mathbb{C}$. A complex valued function is a function from $\mathbb{C}$ to $\mathbb{C}$ of the form

$$f(z) = f(x + iy) = u(x, y) + iv(x, y),$$

with $u(x, y)$, $v(x, y)$ real valued functions. The definitions for continuity and differentiability for a complex valued function are the same as for a real valued function, i.e. $f(z)$ is differentiable at z_0 iff

$$\lim_{\Delta z \to 0} \frac{f(z_0 + \Delta z) - f(z_0)}{\Delta z}$$

exists, if so, the limit is defined as $f'(z_0)$. Over the real number line there are only two directions from which to approach a point – from the left or from the right. The complex line corresponds to a real plane (plot the two real components of $z = x + iy$), so we can approach a point z_0 from many different directions; the limit above exists iff it takes the same value for all possible directions of approach to z_0. The famous Cauchy–Riemann equations tell us that we don't have to check every single direction; it suffices to check the limit along a horizontal and vertical ray:

Theorem B.1.1 (Cauchy–Riemann equations). *Let $f(x + iy) = u(x, y) + iv(x, y)$ and $z_0 = x_0 + iy_0$. If u_x, u_y, v_x, v_y exist and are continuous at*

(x_0, y_0), *then* $f'(z_0)$ *exists iff*

$$u_x = v_y \text{ and } v_x = -u_y.$$

Proof. Write $z_0 = x_0 + iy_0$. First, let $\Delta z = \Delta x + i\,\Delta y$ tend to 0 through values with $\Delta y = 0$. Then the limit above is

$$\lim_{\Delta z \to 0} \frac{f(z_0 + \Delta z) - f(z_0)}{\Delta z} = \lim_{\Delta x \to 0} \frac{u(x_0 + \Delta x, y_0) - u(x_0, y_0)}{\Delta x}$$
$$+ i\frac{v(x_0 + \Delta x, y_0) - v(x_0, y_0)}{\Delta x} = u_x + iv_x.$$

On the other hand, let Δz tend to 0 through values with $\Delta x = 0$. Then we obtain

$$\lim_{\Delta z \to 0} \frac{f(z_0 + \Delta z) - f(z_0)}{\Delta z} = \lim_{\Delta y \to 0} \frac{u(x_0, y_0 + \Delta y) - u(x_0, y_0)}{i\,\Delta y}$$
$$+ i\frac{v(x_0, y_0 + \Delta y) - v(x_0, y_0)}{i\,\Delta y} = -iu_y + v_y.$$

Obviously these must agree if f is differentiable, which shows that the Cauchy–Riemann equations are necessary. We leave sufficiency for the reader to prove or look up. □

When a function $f(z)$ is differentiable in an open neighborhood of a point, then we say that f is *holomorphic* at the point.

B.2 Green's Theorem

Surprisingly, most of the important results in one complex variable are easy consequences of Green's Theorem, which you saw back in vector calculus. Since that may have been awhile ago, we provide a refresher; first let's recall how to do line integrals. Suppose C is a parametric curve; for example consider the plane curve given by $C = (t^2, t^3)$, $t \in [1, 2]$. Notice that the curve is *oriented*: as t goes from 1 to 2, a particle on the curve moves from $(1, 1)$ and ends at $(4, 8)$ (the notation $-C$ means traverse C in the opposite direction).

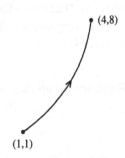

Evaluating a function $g(x, y)$ along C is easy: we just plug in t^2 for x and t^3 for y. But we can't just multiply $g(x, y)$ by dt – to integrate, we want to sum up the function times a little element of arclength ds, which (draw a picture!) is given by

$$\sqrt{\left(\frac{dx}{dt}\right)^2 + \left(\frac{dy}{dt}\right)^2}\, dt.$$

For our parameterization, $\frac{dx}{dt} = 2t$, $\frac{dy}{dt} = 3t^2$. If, say, $g(x, y) = x + 2y^2$, then we obtain

$$\int_C g(x, y)\, ds = \int_1^2 \left(t^2 + 2(t^3)^2\right)\sqrt{(2t)^2 + (3t^2)^2}\, dt.$$

Now that we remember how to do line integrals, we recall the key theorem of vector calculus:

Theorem B.2.1 (Green's Theorem). *Let C be a simple closed contour in the plane, oriented counterclockwise. If R denotes the region bounded by C, and $P(x, y)$ and $Q(x, y)$ are continuously differentiable functions, then*

$$\int_C P(x, y)\, dx + Q(x, y)\, dy = \int\int_R (Q_x - P_y)\, dx\, dy.$$

Proof. A simple closed contour is nothing more than a smooth closed curve. We prove Green's Theorem for the contour $C = C_1 \cup C_2$ pictured below. The general case is easily proved by splitting up the region bounded by C into a bunch of pieces of this form. Orient C counterclockwise, and let C_1 denote the "bottom" curve, and C_2 denote the "top" curve.

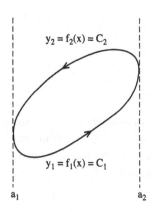

We have:

$$\int_{a_1}^{a_2} \int_{y_1}^{y_2} \frac{\partial P}{\partial y} \, dy \, dx = \int_{a_1}^{a_2} (P(x, f_2(x)) - P(x, f_1(x))) \, dx$$

$$= \int_{-C_2} P \, dx - \int_{C_1} P \, dx = \int_C -P.$$

Almost the same formula holds for Q_x – work through it to see why the signs differ, and put the pieces together to conclude the proof. $\square$

Exercise B.2.2. Let C be the unit circle, parameterized by $x = cos(t)$, $y = sin(t)$, $t \in [0..2\pi]$. Evaluate

$$\int_C xy \, dx + y^2 \, dy$$

directly (using the parameterization), and using Green's Theorem. $\Diamond$

B.3 Cauchy's Theorem

We now return to the complex case. A surprising and easy consequence of Green's Theorem and the Cauchy–Riemann equations is

Theorem B.3.1 (Cauchy's Theorem). *If $f(z)$ is holomorphic within and on a simple closed contour C, then*

$$\int_C f(z) \, dz = 0.$$

Proof. Write $f(z) = u(x, y) + iv(x, y)$ and $dz = dx + i \, dy$. Multiplying out,

$$f(z) \, dz = u(x, y) \, dx - v(x, y) \, dy + i(v(x, y) \, dx + u(x, y) \, dy).$$

So

$$\int_C f(z) \, dz = \int_C u(x, y) \, dx - v(x, y) \, dy + i \int_C v(x, y) \, dx + u(x, y) \, dy.$$

By Green's Theorem, this is

$$\int \int_R -v_x - u_y + i \int \int_R u_x - v_y.$$

But the Cauchy–Riemann equations say these integrals are both zero! Historical note: Cauchy proved this for f' continuous; Goursat showed that the hypothesis f' continuous is unnecessary, so this is sometimes referred to as the Cauchy–Goursat Theorem. $\square$

The Cauchy integral formula says that if C is a simple closed contour and f is holomorphic within and on C, then the value of f at a point interior to C is determined by the values f takes on C. Henceforth, all contours will be oriented *positively*, which corresponds to the choice of counterclockwise orientation of the boundary in Green's Theorem.

Theorem B.3.2 (Cauchy Integral formula). *If $f(z)$ is holomorphic within and on a simple closed contour C, then for any z_0 in the interior of the region bounded by C,*

$$f(z_0) = \frac{1}{2\pi i} \int_C \frac{f(z)\,dz}{z - z_0}.$$

Proof. By parameterizing a little circle C_0 of radius $R = |z - z_0|$ as $z = z_0 + Re^{i\theta}$, we get $dz = iRe^{i\theta}\,d\theta$. A quick computation shows that

$$\int_{C_0} \frac{dz}{z - z_0} = 2\pi i.$$

Now split the annulus as below:

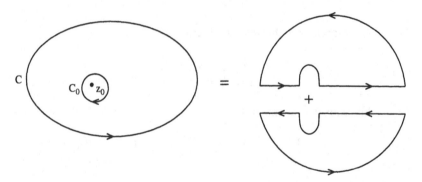

By Cauchy's Theorem,

$$\int_C \frac{f(z)\,dz}{z - z_0} - \int_{C_0} \frac{f(z)\,dz}{z - z_0} = 0,$$

i.e.

$$\int_C \frac{f(z)\,dz}{z - z_0} = \int_{C_0} \frac{f(z)\,dz}{z - z_0}.$$

Now we simply subtract a judiciously chosen quantity from both sides:

$$f(z_0)2\pi i = f(z_0) \int_{C_0} \frac{dz}{z - z_0},$$

obtaining

$$\int_C \frac{f(z)dz}{z - z_0} - f(z_0)2\pi i = \int_{C_0} \frac{f(z) - f(z_0)}{z - z_0}\, dz.$$

Since f is holomorphic, as we shrink C_0 the right hand side goes to zero, done. $\square$

If $f(z)$ is holomorphic within and on a simple closed contour C, then the Cauchy integral formula tells us that for any z in the interior

$$f(z) = \frac{1}{2\pi i} \int_C \frac{f(s)\, ds}{s - z}.$$

So it follows that

$$\frac{f(z + \Delta z) - f(z)}{\Delta z} = \frac{1}{2\pi i} \int_C \left(\frac{1}{s - z - \Delta z} - \frac{1}{s - z} \right) \frac{f(s)\, ds}{\Delta z}$$

$$= \frac{1}{2\pi i} \int_C \frac{f(s)\, ds}{(s - z - \Delta z)(s - z)}.$$

As Δz approaches zero, this quantity is equal to

$$\frac{1}{2\pi i} \int_C \frac{f(s)\, ds}{(s - z)^2}.$$

To see this, consider the integral

$$\int_C \left(\frac{1}{s - z - \Delta z} \cdot \frac{1}{s - z} - \frac{1}{(s - z)^2} \right) f(s)\, ds = \Delta z \int_C \frac{f(s)\, ds}{(s - z - \Delta z)(s - z)^2}.$$

Set $M = \max |f(s)|$ on C, $L = $ length of C. We may assume that $|\Delta z|$ is small enough that $z + \Delta z$ is in the interior of C. Let δ be the smallest distance from z to C. So $|s - z - \Delta z| \geq \big||s - z| - |\Delta z|\big| \geq \delta - |\Delta z|$, and we obtain

$$\left| \Delta z \int_C \frac{f(s)\, ds}{(s - z - \Delta z)(s - z)^2} \right| \leq \frac{|\Delta z| \cdot M \cdot L}{(\delta - |\Delta z|) \cdot \delta^2} \to 0 \quad \text{as} \quad |\Delta z| \to 0,$$

which is the desired result. We have shown

Theorem B.3.3. *If f is holomorphic near a point z, then f' is holomorphic near z, and $f'(z) = \frac{1}{2\pi i} \int_C \frac{f(s)\, ds}{(s - z)^2}$ (C a simple closed contour around z).*

Repeating, we obtain the Cauchy Derivative formula:

$$f^{(n)}(z) = \frac{n!}{2\pi i} \int_C \frac{f(s)\, ds}{(s - z)^{n+1}}.$$

In particular, f is automatically infinitely differentiable.

B.4 Taylor and Laurent Series, Residues

Being holomorphic imposes additional very strong constraints on a complex function. For example, we have

Theorem B.4.1. *If $f(z)$ is holomorphic within a circle of radius R centered at z_0, then $\forall z$ with $|z - z_0| < R$, there is a Taylor series expansion for $f(z)$:*

$$f(z) = \sum_{j=0}^{\infty} \frac{f^{(j)}(z_0)(z - z_0)^j}{j!}.$$

The proof follows from the Cauchy integral and derivative formulas, see [20]. An important variant of the Taylor series is the Laurent series, which allows negative powers in the series above:

Theorem B.4.2. *Let C_0 and C_1 be two positively oriented circles of radius $R_0 < R_1$ centered at z_0. If $f(z)$ is holomorphic in the region $R_0 \leq |z| \leq R_1$, then $\forall z$ such that $R_0 < |z| < R_1$, $f(z)$ has a Laurent expansion, i.e.*

$$f(z) = \sum_{n=0}^{\infty} a_n(z - z_0)^n + \sum_{n=1}^{\infty} \frac{b_n}{(z - z_0)^n}$$

$$a_n = \frac{1}{2\pi i} \int_{C_1} \frac{f(z)\, dz}{(z - z_0)^{n+1}}$$

$$b_n = \frac{1}{2\pi i} \int_{C_0} \frac{f(z)\, dz}{(z - z_0)^{1-n}}.$$

The proof follows the same lines as that for the Taylor series. If $f(z)$ is holomorphic in a neighborhood of a point z_0 but not holomorphic at the point itself, then we say that f has an *isolated singularity* at z_0. An isolated singularity for which the only nonzero b_i in the Laurent series is b_1 is called a *simple pole*; if $b_i = 0$ for $i \gg 0$ then f is *meromorphic* at z_0.

By the previous theorem if f has an isolated singularity, then f has a Laurent series expansion at z_0. The number b_1 which appears in the Laurent series is called the *residue* of f at z_0, and we have the *residue theorem*: for a little simple closed contour C around an isolated singular point z_0,

$$\int_C f(z)\,dz = 2\pi i \cdot b_1.$$

The residue theorem will be important in proving the Riemann–Roch Theorem; we will use the fact that if $f(z)$ has an isolated singularity at z_0, then the residue is a well defined complex number.

Supplemental Reading: A nice elementary complex analysis text is Brown–Churchill [20], which is where I learned most of the material here.

Bibliography

[1] W. Adams, P. Loustaunau, *An introduction to Gröbner bases*, AMS, Providence, RI, 1994.

[2] E. Arbarello, M. Cornalba, P. Griffiths, J. Harris, *Geometry of Algebraic Curves*, Springer–Verlag, Berlin–Heidelberg–New York, 1985.

[3] M. Atiyah, I. MacDonald, *Commutative Algebra*, Addison–Wesley, Reading MA, 1969.

[4] A. Aramova, J. Herzog, T. Hibi, Shifting operations and graded Betti numbers. *J. Algebraic Combin*, **12** (2000), 207–222.

[5] A. Aramova, J. Herzog, T. Hibi, Gotzmann theorems for exterior algebras and combinatorics. *J. Algebra* **191** (1997), 174–211.

[6] S. Balcerzyk and T. Jozefiak, *Commutative Rings. Dimension, Multiplicity and Homological Methods*, Prentice Hall, Englewood Cliffs, NJ, 1989.

[7] D. Bayer, *The division algorithm and the Hilbert scheme*, Thesis, Harvard University, 1982. Available at http://www.math.columbia.edu/~bayer

[8] D. Bayer, H. Charalambous, S. Popescu, Extremal Betti numbers and applications to monomial ideals, *J. Algebra*, **221** (1999), 497–512.

[9] D. Bayer, D. Mumford, What can be computed in algebraic geometry? *Computational algebraic geometry and commutative algebra (Cortona, 1991)*, Sympos. Math., XXXIV, Cambridge University Press, Cambridge–New York, 1993, 1–48.

[10] D. Bayer, I. Peeva, B. Sturmfels, Monomial resolutions. *Math. Res. Lett.* **5** (1998), 31–46.

[11] D. Bayer and M. Stillman, Computation of Hilbert functions. *J. Symbolic Computation*, **14** (1992), 31–50.

[12] D. Bayer and M. Stillman, A theorem on refining division orders by the reverse lexicographic order. *Duke Math. J* **55** (1987), 321–328.

[13] T. Becker, V. Weispfenning, *Gröbner Bases*, Springer–Verlag, Berlin–Heidelberg–New York, 1993.

[14] A. Bigatti, P. Conti, L. Robbiano, C. Traverso, *A "divide and conquer" algorithm for Hilbert-Poincaré series, multiplicity and dimension of monomial ideals*, Lecture Notes in Comput. Sci., 673, Springer, Berlin, 1993.

[15] A. Bigatti, A. Geramita, J. Migliore, Geometric consequences of extremal behavior in a theorem of Macaulay, *Transactions of the AMS* **346** (1994), 203–235.

[16] L. Billera, C. Lee, A proof of the sufficiency of McMullen's conditions for f-vectors of simplicial convex polytopes, *J. Combin. Theory Ser. A* **31** (1981), 237–255.

183

[17] A. Björner, The unimodality conjecture for convex polytopes, *Bulletin of the AMS* **4** (1981), 187–188.

[18] E. Brieskorn, H. Knörrer, *Plane algebraic curves*, Birkhäuser Verlag, Basel, 1986.

[19] M. Brodmann, R. Sharp, *Local Cohomology*, Cambridge University Press, Cambridge–New York, 1998.

[20] J. Brown and R. Churchill, *Complex Variables and Applications*, McGraw–Hill, New York, 1984.

[21] W. Bruns and J. Herzog, *Cohen–Macaulay Rings*, Cambridge University Press, Cambridge–New York, 1998.

[22] D. Buchsbaum, D. Eisenbud, *Gorenstein Ideals of Height* 3, Teubner-Texte zur Math., 48, Teubner, Leipzig, 1982.

[23] D. Cox, J. Little, D. O'Shea, *Ideals, Varieties, and Algorithms*, 2nd edition, Springer–Verlag, Berlin–Heidelberg–New York, 1997.

[24] D. Cox, J. Little, D. O'Shea, *Using Algebraic Geometry*, Springer–Verlag, Berlin–Heidelberg–New York, 1998.

[25] E. Davis, Complete intersections of codimension two in $\mathbb{P}^r$ —The Bezout–Jacobi–Segre theorem revisited, *Rend. Sem. Mat. Univ. Politec. Torino* **43** (1985), 333–353.

[26] W. Decker, F.-O. Schreyer, Computational algebraic geometry today. *Applications of algebraic geometry to coding theory, physics and computation*, NATO Sci. Ser. II Math. Phys. Chem., 36, Kluwer Acad. Publ., Dordrecht, (2001), 65–119.

[27] J. Eagon, V. Reiner, Resolutions of Stanley-Reisner rings and Alexander duality. *J. Pure Appl. Algebra* **130** (1998), 265–275.

[28] D. Eisenbud, *Commutative Algebra with a View Towards Algebraic Geometry*, Springer–Verlag, Berlin–Heidelberg–New York, 1995.

[29] D. Eisenbud, *The Geometry of Syzygies*, Springer-Verlag, to appear, available at: http://www.msri.org/people/staff/de/index.html

[30] D. Eisenbud, S. Goto, Linear free resolutions and minimal multiplicity. *J. Algebra* **88** (1984), 89–133.

[31] D. Eisenbud, M. Green, J. Harris, Cayley-Bacharach theorems and conjectures. *Bull. Amer. Math. Soc.* **33** (1996), 295–324.

[32] D. Eisenbud, D. Grayson, M. Stillman, and B. Sturmfels, *Computations in Algebraic Geometry with Macaulay 2*, Springer-Verlag, 2001.

[33] D. Eisenbud, J. Harris, Curves in projective space. *Seminaire de Mathematiques Superieures* **85** Presses de l'Universite de Montreal, Montreal, Quebec, 1982.

[34] D. Eisenbud, J. Harris, *The geometry of schemes*, Springer-Verlag, New York, 2000.

[35] D. Eisenbud, C. Huneke, W. Vasconcelos, Direct methods for primary decomposition, *Inventiones Math.* **110** (1992), 207–235.

[36] D. Eisenbud and S. Popescu, Gale duality and free resolutions of ideals of points, *Inventiones Math.* **136** (1999), 419–449.

[37] J. Emsalem, A. Iarrobino, Inverse system of a symbolic power I, *J. Algebra* **174** (1995), 1080–1090.

[38] E. G. Evans, P. Griffith, *Syzygies*, London Mathematical Society Lecture Note Series, 106, Cambridge University Press, Cambridge, 1985.

[39] R. Fröberg, An inequality for Hilbert series of graded algebras, *Math. Scand*, **56** (1985), 117–144.

[40] W. Fulton, *Algebraic Curves*, Addison–Wesley, NY, 1989.

[41] W. Fulton, *Algebraic Topology*, Springer–Verlag, Berlin–Heidelberg–New York, 1995.

[42] I. Gelfand, Y. Manin, *Homological algebra*, Springer–Verlag, Berlin–Heidelberg–New York, 1999.

[43] A. Geramita, Inverse Systems of Fat Points: Waring's Problem, Secant Varieties of Veronese Varieties, and Parameter Spaces for Gorenstein Ideals, *Queen's Papers in Pure and Applied Mathematics* **102**, The Curves Seminar, Volume X, (1996), 1–114.

[44] A. Geramita, *Queen's Papers in Pure and Applied Mathematics* **120**, The Curves Seminar, (2000).

[45] D. Grayson and M. Stillman, Macaulay 2, A program for commutative algebra and algebraic geometry, available at: http://www.math.uiuc.edu/Macaulay2

[46] M. Green, *Restrictions of Linear Series to Hyperplanes, and some Results of Macaulay and Gotzmann, Springer LNM 1389*, Algebraic Curves and Projective Geometry, Springer Verlag, (1989).

[47] G. Greuel and G. Pfister, *A Singular Introduction to Commutative Algebra*, Springer-Verlag, 2002.

[48] P. Griffiths, *Introduction to Algebraic Curves*, AMS, Providence RI, 1989.

[49] P. Griffiths and J. Harris, *Principles of Algebraic Geometry*, Wiley and Sons, NY, 1978.

[50] L. Gruson, R. Lazarsfeld, C. Peskine, On a theorem of Castelnuovo, and the equations defining space curves. *Inventiones Mathematicae*, **72** (1983), 491–506.

[51] B. Harbourne, Problems and Progress: A Survey on Fat Points in $\mathbb{P}^2$. Zero-Dimensional Schemes and Applications (Naples, 2000), 85–132. *Queen's Papers in Pure and Appl. Math.*, **123**, Queen's Univ., Kingston, ON, 2002.

[52] J. Harris, *Algebraic Geometry – A First Course*, Springer–Verlag, Berlin–Heidelberg–New York, 1992.

[53] R. Hartshorne, *Algebraic Geometry*, Springer Verlag, Berlin–Heidelberg–New York, 1977.

[54] I. Herstein, *Topics in Algebra*, Wiley and Sons, New York, 1975.

[55] M. Hochster, Cohen-Macaulay rings, combinatorics, and simplicial complexes. Ring theory, II (Proc. Second Conf., Univ. Oklahoma, Norman, Okla., 1975), pp. 171–223. Lecture Notes in Pure and Appl. Math., Vol. 26, Dekker, New York, 1977.

[56] T. Hungerford, *Algebra*, Springer–Verlag, Berlin–Heidelberg–New York, 1974.

[57] A. Iarrobino, V. Kanev, Power sums, Gorenstein algebras, and determinantal loci. Appendix C by Iarrobino and Steven L. Kleiman. *Lecture Notes in Mathematics*, **1721**, Springer-Verlag, Berlin, 1999.

[58] M. Kreuzer, L. Robbiano, *Computational Commutative Algebra I*, Springer-Verlag, 2000.

[59] S. Lang, *Algebra*, Addison-Wesley, 1993.

[60] R. Lazarsfeld, A sharp Castelnuovo bound for smooth surfaces. *Duke Math. J.*, **55** (1987), 423–429.

[61] C. Lee, *Counting the faces of simplicial polytopes*, Thesis, Cornell University, 1981.

[62] A. Lorenzini, The minimal resolution conjecture, *J. Algebra*, **156** (1993), 5–35.

[63] F.H.S. Macaulay, *The algebraic theory of modular systems*, Cambridge University Press, London/New York, 1916.

[64] H. Matsumura, *Commutative Ring Theory*, Cambridge University Press, Cambridge–New York, 1986.

[65] E. Mayr, A. Meyer, The complexity of the word problems for commutative semi-groups and polynomial ideals, *Adv. in Math*, **46** (1982), 305–329.

[66] J. Migliore, R. Miro-Roig, Ideals of general forms and the ubiquity of the Weak Lefschetz property, available at: math.AG/0205133.

[67] J. Migliore, U. Nagel, Reduced arithmetically Gorenstein schemes and simplicial polytopes with maximal Betti numbers, *Adv. in Math*, (to appear), available at: math.AG/0103197.

[68] E. Miller, The Alexander duality functors and local duality with monomial support, *J. Algebra* **231** (2000), 180–234.

[69] R. Miranda, *Algebraic Curves and Riemann Surfaces*, AMS, Providence RI, 1995.

[70] R. Miranda, Linear systems of plane curves, *Notices of the AMS.* **46** 2 (1999), 192–202.

[71] J. Munkres, *Elements of Algebraic Topology*, Benjamin–Cummings, Menlo Park, 1984.

[72] J. Munkres, Topological results in combinatorics. *Michigan Math. J.*, **31** (1984), 113–128.

[73] M. Mustaţă, Graded Betti numbers of general finite subsets of points on projective varieties. *Matematiche (Catania)*, **53** (1998), 53–81.

[74] R. Narasimhan, *Compact Riemann Surfaces*, Birkhäuser Verlag, Basel, 1992.

[75] P. Orlik, H. Terao, *Arrangements of Hyperplanes*, Springer-Verlag, Berlin-Heidelberg-New York, 1992.

[76] K. Pardue, B. Richert, Szygyies of Semi-Regular Sequences, preprint, 2002.

[77] H. Pinkham, A Castelnuovo bound for smooth surfaces. *Invent. Math.* **83** (1986), 321–332.

[78] M. Reid, *Undergraduate Algebraic Geometry*, Cambridge University Press, Cambridge–New York–Sydney, 1988.

[79] G. Reisner, Cohen-Macaulay quotients of polynomial rings, *Advances in Math*, **21** (1976), 30–49.

[80] J. Rotman, *An introduction to algebraic topology*, Springer-Verlag, New York, 1988.

[81] J. Rotman, *Advanced modern algebra*, Prentice Hall, 2002.

[82] I.R. Shafarevich, *Basic algebraic geometry. 1. Varieties in projective space*, Springer-Verlag, Berlin, 1994.

[83] I.R. Shafarevich, *Basic algebraic geometry. 2. Schemes and complex manifolds*, Springer-Verlag, Berlin, 1994.

[84] R. Sharp, *Steps in Commutative Algebra*, Cambridge University Press, Cambridge–New York–Sydney, 2000.

[85] G. Smith, Computing global extension modules, *J. Symbolic Comput*, **29** (2000), 729–746.

[86] K. Smith, L. Kahanpää, P. Kekäläinen, and W. Traves, *An Invitation to Algebraic Geometry*, Springer, 2000.

[87] E. Spanier, *Algebraic Topology*, McGraw–Hill, New York, 1966.

[88] R. Stanley, *Commutative Algebra and Combinatorics*, Birkhauser, Berlin–Heidelberg–New York, 1996.

[89] R. Stanley, The upper bound conjecture and Cohen–Macaulay rings., Studies in Appl. Math. **54** (1975), 135–142.

[90] R. Stanley, The number of faces of a simplicial convex polytope, *Adv. in Math*, **35** (1980), 236–238.

[91] M. Stillman, Methods for computing in algebraic geometry and commutative algebra, *Acta Appl. Math* **21** (1990), 77–103.

[92] B. Sturmfels, *Gröbner bases and convex polytopes*, American Mathematical Society, University Lecture Series 8, 1995.

[93] B. Sturmfels, *Solving Systems of Polynomial Equations*, American Mathematical Society, C.B.M.S. Lecture Notes 97, 2002.

[94] M. Teixidor, Green's conjecture for the generic r-gonal curve of genus $g \geq 3r - 7$, *Duke Math. J*, **111** (2002), 195–222.

[95] W. Vasconcelos, *Computational Methods in Commutative Algebra and Algebraic Geometry*, Springer–Verlag, Berlin–Heidelberg–New York, 1998.

[96] C. Voisin, Green's generic syzygy conjecture for curves of even genus lying on a K3 surface, available at: math.AG/0301359

[97] R. Walker, *Algebraic Curves*, Princeton University Press, 1950.

[98] C. Weibel, *An Introduction to Homological Algebra*, Cambridge University Press, Cambridge, 1994.

[99] J. Weyman, *Cohomology of Vector Bundles and Syzygies*, London Mathematical Society Lecture Note Series, Cambridge University Press, 2003.

[100] G. Ziegler, *Lectures on Polytopes*, Springer–Verlag, Berlin–Heidelberg–New York, 1995.

Index

Printed in the United States
By Bookmasters